Contents

Canadian Marine Fisheries in a Changing and Uncertain World

NRC Monograph Publishing Program

Enquiries: Monograph Publishing Program, NRC Research Press, National Research Council of Canada, Ottawa, Ontario K1A 0R6, Canada

Correct citation for this publication:

deYoung, B., Peterman, R.M., Dobell, A.R., Pinkerton, E., Breton, Y., Charles, A.T., Fogarty, M.J., Munro, G.R., and Taggart, C.T. 1999. *Canadian Marine Fisheries in a Changing and Uncertain World.* Can. Spec. Publ. Fish. Aquat. Sci. 129. 199 p.

Canadian Special Publication of Fisheries and Aquatic Sciences 129

Canadian Marine Fisheries in a Changing and Uncertain World

A report prepared for the
Canadian Global Change Program of the Royal Society of Canada

Panel Members and Authors

Brad deYoung (*co-chair*), *Department of Physics and Physical Oceanography, Memorial University of Newfoundland, St. John's, Newfoundland A1B 3X7, Canada.*

Randall M. Peterman (*co-chair*), *School of Resource and Environmental Management, Simon Fraser University, Burnaby, British Columbia V5A 1S6, Canada.*

A. Rod Dobell, *School of Public Administration, University of Victoria, Victoria, British Columbia, V8W 2Y2, Canada.*

Evelyn Pinkerton, *School of Resource and Environmental Management, Simon Fraser University, Burnaby, British Columbia V5A 1S6, Canada.*

Yvan Breton, *Department of Anthropology, Université Laval, Ste-Foy, Québec G1K 7P4, Canada.*

Anthony T. Charles, *Department of Finance and Management Science, Saint Mary's University, Halifax, Nova Scotia B3H 3C3, Canada.*

Michael J. Fogarty, *Chesapeake Biological Laboratory, University of Maryland, Solomons, Maryland 20688, U.S.A.*

Gordon R. Munro, *Department of Economics, University of British Columbia, Vancouver, British Columbia V6T 1Z1, Canada.*

Christopher T. Taggart, *Department of Oceanography, Dalhousie University, Halifax, Nova Scotia B3H 4J1, Canada.*

NRC·CNRC
NRC Research Press
Ottawa 1999

Printed in Canada on acid-free paper. ♾

ISBN 0-660-17638-6
ISSN 0706-6481
NRC No. 42732

Canadian Cataloguing in Publication Data

Canadian Marine Fisheries in a Changing and Uncertain World

(Canadian special publication of fisheries and aquatic sciences,
ISSN 0706-6481; no. 129)
"A publication of the National Research Council of Canada
Monograph Publishing Program"
Includes an abstract in French.
Includes bibliographical references.
ISBN 0-660-17638-6

1. Sustainable fisheries — Canada. 2. Fisheries — Canada.
3. Fishery management — Canada.
I. deYoung, Brad, 1956– . II. Canadian Global Change Program.
III. National Research Council Canada. IV. Series.

SH223.C32 1999 333.95'6'0971 C98-980356-2

The views expressed are those of the authors and not necessarily those of the National Research Council of Canada.

Abstract

In recent years, Canadians have become increasingly aware of the dynamic and changing nature of fishery systems, which include not only the fish and their environment, but also people and their associated social and economic institutions and communities. As a result of the challenges created by our constantly changing fisheries, the Canadian Global Change Program of the Royal Society of Canada formed an interdisciplinary Fisheries Panel of nine people in 1996. The panel's mandate was to write an authoritative and comprehensive review of the implications of physical, biological, economic, and socio-political changes for Canadian marine fisheries and to present options for how to deal with those changes.

This report outlines a vision for future sustainable marine fisheries for Canada and reviews the state of Canadian marine fishery systems. It also examines the complexities and major processes of change and variability in fishery systems, which lead to uncertainties and create risks, i.e., biological risks for fish populations, economic risks for those in industry, and social and economic risks for people in coastal communities who rely on renewable aquatic resources. These risks have important implications not only for how management agencies should operate, but also for how industry, fishermen, and coastal fishing-dependent communities should plan and act. The last part of the report therefore emphasizes the features of an effective social response to the challenge of managing these risks: stewardship and conservation, participation and cooperation, and compliance and accountability. We develop several guiding principles that, in the presence of change, uncertainty, and risks, can help attain the overall goals of biologically productive aquatic systems, economically viable fishing industries, and sustainable fishing-dependent communities. These principles include (*a*) incorporating analysis of structural and dynamic complexities of fishery systems into decision-making; (*b*) incorporating explicit analysis of change, uncertainties, and risk into decision-making; (*c*) promoting and conserving biological, economic, and social diversity; (*d*) collecting, analyzing and openly communicating data and information; (*e*) estimating and documenting the social and ecological consequences of decisions and actions; and (*f*) clearly defining roles, rights, and responsibilities of all fishery participants to align their interests with the overall objectives of sustainability. Finally, we identify several strategies that can be used by management agencies, industry, fishermen, and coastal fishing-dependent communities to follow these principles. These strategies include implementing (*a*) a precautionary approach, (*b*) risk-assessment and risk-management procedures, (*c*) ecosystem-based management and promoting diversity, (*d*) measures to ensure adequate collection and communication of information, (*e*) institutional reform and promoting compliance, and (*f*) informed social decisions through appropriate pricing, accounting, reporting, and charging. The benefits and drawbacks of these different strategies are assessed and examples are provided of their application in fishery systems.

Résumé

Ces dernières années, les Canadiens ont de plus en plus pris conscience de la nature dynamique et changeante des systèmes halieutiques, qui n'englobent pas uniquement les poissons et leur environnement, mais aussi des gens, des communautés et les institutions économiques et sociales qui y sont associées. Comme l'évolution constante de nos pêches pose de difficiles défis, le Programme canadien des changements à l'échelle du globe de la Société royale du Canada a constitué en 1996 un comité pluridisciplinaire des pêches composé de neuf personnes. Celui-ci a reçu pour mandat de préparer une analyse exhaustive et faisant autorité des répercussions des changements physiques, biologiques, économiques et sociopolitiques qui touchent les pêches en mer du Canada et de décrire les options qui permettent d'y faire face.

Ce rapport présente une vision de ce qui peut rendre les pêches en mer durables au Canada. Nous passons tout d'abord en revue l'état des systèmes halieutiques marins canadiens. Nous examinons ensuite les complexités et les principaux mécanismes de changement et de variabilité de ces systèmes, des mécanismes qui engendrent des incertitudes et des risques — des risques biologiques pour les populations de poissons, des risques économiques pour les membres de l'industrie et des risques économiques et sociaux pour les habitants des collectivités côtières qui dépendent des ressources aquatiques renouvelables. Ces risques ont des conséquences importantes non seulement pour le fonctionnement des agences de gestion, mais aussi en ce qui concerne la manière dont l'industrie, les pêcheurs et les collectivités côtières qui dépendent de la pêche devraient planifier leurs activités et conduire celles-ci. La dernière partie du rapport fait donc ressortir les caractéristiques d'une réaction sociale efficace au défi que pose la gestion de ces risques : saine intendance environnementale et conservation, participation et coopération, obligation de rendre des comptes et respect de la réglementation. Nous formulons plusieurs principes directeurs qui, dans des situations marquées par des changements, de l'incertitude et des risques, peuvent contribuer à la réalisation des objectifs d'ensemble de productivité biologique des systèmes aquatiques, de viabilité économique des industries de la pêche et de durabilité des communautés qui dépendent de la pêche. Ces principes sont les suivants : (*a*) intégration dans le processus décisionnel d'une analyse de la complexité structurelle et dynamique des systèmes halieutiques; (*b*) intégration dans ce processus d'une analyse explicite des changements, des incertitudes et des risques; (*c*) promotion et conservation de la diversité biologique, économique et sociale; (*d*) collecte, analyse et libre diffusion des données et des informations; (*e*) estimation et documentation des conséquences écologiques et sociales des décisions et des actes; et (*f*) définition claire des rôles, des droits et des responsabilités de tous ceux qui oeuvrent dans le domaine de la pêche afin d'aligner leurs intérêts sur l'objectif d'ensemble de durabilité. Enfin, nous présentons plusieurs stratégies grâce auxquelles les agences de gestion, l'industrie, les pêcheurs et les communautés côtières qui dépendent de la pêche peuvent mettre ces principes en pratique. Ces stratégies consistent en : (*a*) une approche prudente, (*b*) des mécanismes d'évaluation et de gestion des risques, (*c*) une gestion écosystémique et la promotion de la diversité, (*d*) des mesures garantissant une collecte et une diffusion adéquates de l'information; (*e*) une réforme institutionnelle et la promotion du respect de la réglementation; et (*f*) des mesures visant à ce que les décisions sociales soient éclairées par des mécanismes appropriés d'établissement des prix, de comptabilisation, de préparation de rapports et d'attribution des coûts. Les avantages et les inconvénients de ces stratégies sont évaluaté également en fournissant des exemples de leur mise en oeuvre dans des systèmes halieutiques.

Acknowledgments

The panel is extremely grateful for the significant support and encouragement provided by the Canadian Global Change Programme (CGCP) of the Royal Society of Canada, particularly through the efforts of Jeff Watson and Jim Bottomley in the CGCP office. Additional financial support was provided by the Canada Department of Fisheries and Oceans and the Canada Department of Environment. The panel also received useful advice from George Needler and Rosemary Ommer, members of the Research and Policy Committee of the CGCP who served as outside observers for the panel work. Dr. Ommer went above and beyond the normal expectations of such observers and provided significant input and advice to the panel. We also received considerable support from Dr. John Kearney of St. Francis Xavier University, who assisted us with information and analysis in Chapters 2, 3, and 6 by writing most of the section on the Scotia-Fundy Boards in Chapter 6. Dr. Michael Sissenwine also provided useful guidance when the panel planned its future direction at its first meeting. We are also grateful to Calvin Peters and Jeff Grout of Simon Fraser University, who were research assistants on the project, arranging logistics and taking minutes for several panel meetings, as well as helping out with the compilation of information and writing. As well, the panel received considerable assistance from F. Davidson, K. Forward, M. Jones, A. Pickle, M. Wright, and numerous academic, government, and industrial colleagues (too many to name) in compiling information reviewed in this report. We are also very grateful to the 19 anonymous referees drawn from academia, government, and industry who reviewed the two drafts of this report. Their constructive criticisms and insights led to significant improvements in the document.

Biographies of Panel Members and Authors

Brad deYoung

Brad deYoung is an Associate Professor in the Department of Physics and Physical Oceanography at Memorial University in St. John's where he teaches physical oceanography, physics, ocean ecology, and environmental studies. He is a physical oceanographer who studies the coastal ocean with a focus on ecosystem dynamics, exploring the influence of the physical environment on marine organisms. He presently co-chairs the GLOBEC Canada science program. e-mail: bdeyoung@physics.mun.ca

Randall M. Peterman

Randall Peterman is a Professor in the School of Resource and Environmental Management at Simon Fraser University in Burnaby, British Columbia. He is a fish population ecologist who specializes in applying quantitative methods for dealing with uncertainty in order to improve fisheries management. He and his graduate students have received awards for their research from the Canadian Conference for Fisheries Research (J.C. Stevenson Award), the American Fisheries Society, and the American Institute of Fishery Research Biologists. e-mail: peterman@sfu.ca

A. Rod Dobell

Rod Dobell is Professor of Public Policy at the University of Victoria in Victoria. Early academic specialization in mathematical economics and optimal control theory has been tempered by more recent conversions to the new institutionalism and political philosophy; his field of study is now "ecological socio-political economy." He has held appointments as Professor of Political Economy at the University of Toronto; Deputy Secretary, Treasury Board Secretariat, in the Government of Canada; Director, General Economics Branch at the OECD; Director, School of Public Administration, University of Victoria; and President, The Institute for Research on Public Policy. His current teaching and research interests are in the philosophy of administration and implementation of public policy in the "full world." e-mail: rdobell@uvic.ca

Evelyn Pinkerton

Evelyn Pinkerton is an Associate Professor in the School of Resource and Environmental Management at Simon Fraser University in Burnaby. With a background in cultural ecology and maritime anthropology, her research in Canada, the Pacific Northwest, and Alaska has focused on fishing communities' relationship to adjacent marine resources and the specification of conditions under which successful co-management relationships between communities and governments can arise and persist. e-mail: evelyn_pinkerton@sfu.ca

Yvan Breton

Yvan Breton is a Professor of Anthropology in the Faculty of Social Sciences at Laval University in Québec. He has interests in Maritime and Economic Anthropology and peasant studies. He has conducted research in various parts of Latin American and Africa and is presently involved as an IDRC collaborator in a coastal zone management program in the Caribbean. e-mail: Yvan.Breton@ant.ulaval.ca

Anthony T. Charles

Anthony Charles is a Professor at Saint Mary's University in Halifax, with a joint appointment in Management Science and Environmental Studies. He specializes in interdisciplinary analysis of fisheries, aquaculture, and coastal management problems, with a focus on uncertainty and sustainability, and drawing on methods ranging from natural resource modelling to policy analysis. He served for five years on Atlantic Canada's Fisheries Resource Conservation Council, worked on a variety of international projects for the Food and Agriculture Organization (FAO) and other organizations, and is author of the book *Sustainable Fishery Systems*. e-mail: t.charles@stmarys.ca

Michael J. Fogarty

Michael Fogarty is a fisheries scientist specializing in the population and community dynamics of exploited marine populations. Dr. Fogarty has held positions at the Northeast Fisheries Science Center of the U.S. National Marine Fisheries Service in Woods Hole, Mass., specializing in stock assessment and ecological studies of exploited communities and at the University of Maryland Center for Environmental Science, Chesapeake Biological Laboratory, where he taught courses in fishery science. He currently chairs the U.S. GLOBEC Program and is a member of the International Steering Committee for the Global Ocean Observing System. e-mail: mfogarty@whsun1.wh.whoi.edu

Gordon R. Munro

Gordon Munro is a Professor of Economics at the University of British Columbia in Vancouver. He teaches the economics of natural resource management, along with several other fields. His research in fisheries, which has been both theoretical and applied, has been on various aspects of the economics of fisheries management. The resource management problems, both domestic and international, arising from Extended Fisheries Jurisdiction and the New International Law of the Sea have provided the primary focus for his research activities. e-mail: munro@econ.ubc.ca

Christopher T. Taggart

Christopher Taggart is an Associate Professor of Oceanography at Dalhousie University in Halifax where he teaches fisheries oceanography in the Faculty of Science as well as marine science and technology in the Marine Affairs Programme to Master of Marine Management candidates. His research in fisheries is both fundamental and applied, field and lab, covering physical, ecological, biochemical and genetic aspects of early life history and recruitment in fish and fish population structure and distribution. e-mail: chris.taggart@dal.ca

Chapter 1. Introduction

In recent years, Canadians have heard much about the predicament of their marine fisheries. There have been biological problems such as large decreases in abundance (e.g., the Northern cod on the east coast) and various economic and social issues raised by reduced profitability or the allocation of fish among different user groups. Harvesting pressure has grown because of the increased human population, greater scale of economic and industrial activity, and more sophisticated fishing technology. Growing human settlements and activity in coastal areas have led to direct loss of habitat that is important for spawning or rearing of fish. In addition, there is now increased concern about the effects of changes in the environment, such as those caused by global climatic changes. It thus appears that serious challenges face not just fish, but also people, including fishermen, plant workers, marketers, scientists, and managers. Furthermore, it is possible that our intervention through fishing has created lasting changes. The consequences of such changes are only partly known, but they likely affect fish, other aquatic species, the ocean ecosystem, and ultimately those who depend on fish.

Canadians are concerned about these warning signals because we value fisheries in many ways. Fisheries have been a source of economic wealth to those who work at sea and in fisheries-related industries and those who live in fishing-dependent communities. Furthermore, the heritage of many Canadians is strongly linked to the ocean. There is indeed a cultural and social imperative that gives value to the ocean, the fish, and other organisms in it — witness the growth of eco-tourism and the developing awareness of marine mammals. Canadians also increasingly recognize their stewardship role in protecting marine resources for present and future generations.

Although challenging, the present situation does provide us with opportunities. The recent well-publicized difficulties concerning groundfish on the Atlantic coast and salmon on the Pacific coast have brought fisheries issues to the forefront of public debate in Canada. The growing awareness and public discussion provide us with an opportunity to chart new directions in fisheries policy. This focus on fisheries problems must also be placed in the context of the recent summits on the environment and global change, where there has been widespread public recognition of the need for a collective response.

Approach

It was in this context that the Canadian Global Change Program of the Royal Society of Canada formed a Fisheries Panel of nine people. The panel met at intervals between January 1996 and January 1998 to write this report. The terms of reference for the panel were to write an authoritative and comprehensive review of the implications of physical, biological, economic, and socio-political changes for Canadian marine fisheries and to present options for how to deal with those changes. This focus on change, rather than on fisheries management per se, arose from the increasing awareness of the dynamic, not static, nature of fisheries systems. Here, we use the term "fisheries system" to include fish in their environment, as well as the associated humans and their social and economic structures.

Furthermore, we are now mindful that natural and social systems are complex in a fashion that may fundamentally limit our knowledge of them and our ability to make forecasts of their behavior. Structural complexity has important implications for Canadian fisheries. For instance, change and variability are common to most complex systems in the natural world. Variability implies some fluctuations around a fixed state. Change, which may be gradual or abrupt, refers to movement from one state or condition to another. Variability might refer to the year-to-year changes in catch or abundance, while change could be an increased dominance in a marine community of a previously minor fish species. Of course, both change and variability can occur together and they are not limited to fish populations or the ocean. Economic and social structures also exhibit change and variability, either in response to the underlying natural system or as a result of their own dynamics.

Our knowledge of fishery systems, which is limited to just a few components of their structure, is dwarfed by our ignorance of their overall functioning. One fundamental outcome of change and variability in complex systems is uncertainty, which we discuss in detail in this report. Such uncertainty has important implications for scientists (who must provide analyses), for managers (who must develop policies and plans), for fishermen and industry (who must decide on investment strategies), and for communities (which must survive in and adapt to these changing conditions). In the absence of certainty, the best decision in any particular situation is not easy to identify, partly because risks are difficult to estimate. In the face of these problems, we must be careful not to become paralyzed by our incomplete knowledge. For one thing, our understanding of the natural world is advancing rapidly. Furthermore, very complicated behavior can be driven by relatively few key components of any system and we may well be able to understand components of fishery systems without having to understand all the details.

While the panel focused on Canadian fisheries, it placed them in a broad global context for comparison and drew upon experience in fisheries elsewhere. Given our mandate and the goals — to address the implications of physical, biological, economic, and socio-political changes for Canadian marine fisheries — the panel required an interdisciplinary approach. Its members therefore came from diverse disciplines, including anthropology, economics, fisheries biology, oceanography, and public policy analysis. Such an integrated interdisciplinary analysis of fishery systems is surprisingly rare.

In the context of the many changes that face fishery systems, our work focused on how to attain three goals: a biologically productive aquatic system, economically viable fishing industries, and sustainable fishing-dependent communities. For brevity, we refer to these as the biological, economic, and social components of sustainability. To achieve our goals, the panel drew upon current research in the natural and social sciences, as well as research into appropriate social and institutional responses to uncertainty and risks.

Results

The report begins with an overview of the recent history and current state of Canadian marine fisheries from the perspective of the natural and social sciences (Chapter 2). We show that the current state of many Canadian marine fisheries is not unusual compared to other fisheries in the world. These other fisheries have had similar problems and Canadians can learn from their experience.

We then examine mechanisms of change and variability in the physical, biological, economic, and social components of fishery systems (Chapter 3). We first describe variability in the biological components and present evidence that variability in the physical environment can create variability in marine fish populations. To help the reader understand these processes, we provide some background information on fish in the marine environment, focusing on cod and salmon.

We then explore how variability and change in social and economic processes and the evolution of human institutions influence fishery systems. We examine basic problems of overinvestment in harvesting capacity and underinvestment in conservation and habitat protection. One key element of the present situation is that we have the ability to catch and process far more fish than can be sustainably harvested. Overcapacity is associated with large investments in fishing and processing by people who exert pressure on managers to take risks that are unacceptable to sustainability of the resource. Overcapacity has developed over the past several decades as a result of a number of different factors, including a growing human population and demand for food, developing technology, and social policy that has supported fisheries through government subsidization.

Underlying the overall orientation of this report is the idea that the scale and nature of human interventions must be brought within the capacity of the natural system to accommodate them. In particular, this means that the stress imposed by human activity on marine ecosystems through harvesting must be brought within acceptable limits of the vulnerability of those systems. The current situation of overcapacity is in the long-term interest of no one, carrying as it does the chance of irreversible declines in fish populations, economic collapse, and social collapse for fishing-dependent communities. Why does this situation persist in spite of the growing recognition of many of the symptoms and consequences of the problems? In part it is because of lack of certainty about what to do next. Uncertainty and unavoidable limits to our understanding of fishery systems arise from the complex dynamics of the ecosystem, their changing nature, and the web of incentives and stimuli influencing individual behavioral responses of fishermen.

Thus, three major challenges to sustainability are complexity, change, and uncertainty, which in turn create risks i.e., biological risks for fish populations and economic risks for those in the industry who rely on productive aquatic systems. Chapter 4 elaborates on sources of uncertainty and risks. It sets the stage for later discussions of their implications, not only for actions of management agencies, but also for actions of industry, fishermen, and coastal fishing-dependent communities. The last part of the report therefore emphasizes the features of an effective social response to the challenge of managing these risks: stewardship and conservation, participation and cooperation, and compliance and accountability.

By taking different disciplinary perspectives, we develop new insights into Canadian and global fisheries and derive some fundamental guiding principles (Chapter 5). When followed, these principles can help attain the overall goals of a biologically productive aquatic system, economically viable fishing industries, and sustainable fishing-dependent communities. These principles touch on the following issues. All people involved in making decisions in fishery systems (including managers in harvesting and processing companies, as well as government managers) should incorporate into decision making the complexities, uncertainties, and risks inherent in fisheries. Furthermore, actions should be taken that promote biological diversity in aquatic systems and diversity of employment opportunities in human coastal communities. In addition, information on physical, biological, economic, and social components of fisheries should be collected, analyzed, and communicated more broadly. This will help meet another principle, to more fully estimate and take into account the social and ecological consequences of decisions and actions. For instance, we need to include in evaluations of proposed actions the effects of by-catch and discarding, as well as the non-market value of healthy aquatic systems. A further principle is that people must clearly understand their roles, rights, and attendant social and environmental responsibilities. Within the changing and uncertain world that we describe, the commitment to sustainable use may demand constraints on the exercise of rights. Although such constraints have been recognized in the past, social and economic forces have usually overwhelmed the principle of sustainability.

In Chapter 6, we suggest a range of strategies that can be implemented by individuals, communities, industry, and governments in order to follow these guiding principles. These strategies will help attain productive, long-term sustainable harvests of Canadian fish stocks, and the associated economic and social benefits for industry, fishing communities, and governments. We assess the benefits and drawbacks of these different strategies and provide examples where they have been applied. We emphasize the importance of institutional design and reform to reorient the structures within which there will be widely participatory policy formation and cooperative implementation. We consider various means to attain the goals of biological, economic, and social sustainability. For instance, how should incentives be set up so that individual behavior is aligned with the larger goal of sustainability? We have to "get the incentives clear, and get them right." Should the incentives be created through direct governmental regulation, economic instruments, community quotas, self-governing institutions at regional or community level, or other mechanisms? We review some of these options without trying to be prescriptive. Whatever the decision (and there may be no single solution), the need is clear for administratively feasible, compliance-friendly fisheries management. Our collective and individual goals in exploiting fisheries systems must somehow be brought into line with the global requirements for sustainability.

Other strategies that we propose include taking precautionary and integrated, ecosystem-based approaches. In addition, we need to carefully estimate the various types of risks (biological, economic, and social) associated with planned actions and use those estimates to make better decisions in all components of fisheries. Those components include not just government regulations, but also investing and harvesting decisions by fishermen, fishery unions, processors, and local communities. In the end, behavior on shore and at sea must change in order to promote sustainable marine ecosystems and sustainable fisheries. We must fully accept our responsibility of stewardship for the oceans' resources.

However, we must not be fooled into believing that putting in place a few new policies or adding some new institutions will solve our problems. Fundamental change is required. For example, it is also essential that we begin to implement the notion of "full-cost pricing," so that so-called "side effects" are accounted for before action is taken and not just noted as regrettable outcomes afterwards. Furthermore, we need effective processes and forums in which to hear, interpret, and debate all the available advice and information on fisheries, which is essential to the success of any policy development or implementation. Political action must then develop around the consensus built out of such an open process. Our report is part of the information sharing process, an attempt by a small group to integrate available natural and social science analyses of fisheries.

We explore some strategies for possible institutional design, a subject that quite clearly is one of implementation. It might be argued that this is beyond the expertise of the panel, however, it is abundantly clear that problems in fisheries extend from the biological through the natural environment to the social and institutional. Our integrated and interdisciplinary analysis compels us to follow our arguments about a need for social change with arguments about a need for institutional change.

Chapter 7 summarizes the report and indicates the extent to which the strategies described in Chapter 6 are currently being implemented in Canadian marine fisheries. It therefore points out steps that all Canadians involved in fisheries will need to take in the future.

Outlook

This report outlines a vision for future sustainable fisheries in Canada founded upon the fundamental conviction that biological, economic, and social sustainability are crucial to ensure continued ecosystem integrity, which is the central and overriding objective. Obviously, without fish there cannot be generation of fishing-dependent wealth and without that human communities dependent on fishing cannot be viable. Ecosystem integrity therefore serves to integrate the three components of sustainability. Whether one is concerned about the health of marine ecosystems for their own sake or concerned only about the long-term welfare of people relying on the social and economic benefits derived from those ecosystems, the task is the same. It is to maintain fully functioning ecosystems that are able to cope with natural variability, change of all types, and human intervention. Diversity in both the natural and social systems is crucial in determining the capacity to respond to change. In our analysis of ecosystems, we explicitly include humans because we are quite obviously part of ecosystems and have a measurably growing impact.

This report is intended primarily for those with a direct involvement in fisheries but should also be useful to anyone interested in learning about fisheries issues. Because this report integrates information from each of several disciplines, some readers may find that they are already very familiar with some sections of Chapters 2 and 3. We hope, however, that other parts of the report will be informative and we therefore encourage readers to work through the entire document, passing lightly over sections with which they are familiar, and concentrating on the remainder of the report.

There are necessarily many aspects of fisheries that are not addressed here. First, the panel's terms of reference were to explore the effects on fisheries of processes of change. Thus, some direct impacts of humans on the environment, such as the influence of carbon dioxide emissions on climatic change, receive little attention, although we do address the effect of climatic change on fisheries. Second, our focus is on capture fisheries, not aquaculture. We recognize, however, the significance of growing supplies of farmed fish on world markets as an important global force leading to the need for adaptation in Canadian fisheries and communities.

Finally, this report does not aim to correct all the ills of Canadian marine fisheries. Instead, the goal is to describe the origins of some ongoing changes and their implications for fisheries. The report might, therefore, offer some guidance for all those involved in fisheries: commercial harvesters, sports anglers, plant workers, scientists, unions, managers, and people living in coastal communities. Many fisheries are in a bleak, but perhaps not yet catastrophic, state. From our analysis, we conclude that acceptance and application of our fundamental principles and their associated strategies of implementation should improve the present situation. The following chapters attempt to make that case.

Chapter 2. Marine Fisheries: Canada in the Global Context

Introduction

The first step in our exploration of Canadian marine fisheries is to examine their state and recent history, placing Canadian marine fisheries in a global context. Our goal here is not to present a simple status report but to discuss and explain what we understand about the fisheries and why they are in their observed state. This introductory analysis is essential to appreciate the arguments that follow in Chapters 3 and 4 where we analyze the nature of change and the measurable degrees of uncertainty and risk that are common to all fishery systems.

In this chapter we consider all aspects of fisheries: social, economic, and scientific. We begin with a brief review of basic fish biology, describing typical life histories of the major species harvested in Canadian waters. The focus is on salmon on the west coast fishery and cod on the east coast fishery because they are prime examples of species with significant ecological and socio-economic value. These two species also demonstrate two different life histories of fish and illustrate the adaptation of different species to their aquatic environments. Summaries of the current status of major Canadian fish stocks reveal that the contemporary nature of the Canadian fishery system is neither unique nor unexpected from a global perspective. The reasons for the generally depressed status of many stocks in Canada and throughout the world are then explored through an examination of fishing practices. This leads to an overview of the changing economic, socio-political, and management aspects of the Canadian fishery.

What are marine fishery resources?

Marine fishery resources are loosely defined as a collection of populations of wild animal species (and in a few cases, plants such as kelp) in the ocean that are renewable and subject to human exploitation. The animals most frequently harvested fall into two major categories: (1) fish, which are classified as vertebrates, and (2) shellfish (e.g., scallops), crustaceans (e.g., lobsters, prawns, or shrimp), and cephalopods (e.g., octopus, squid) known as invertebrates. Natural variability in the environment influences these cold-blooded organisms. Thus, even in the absence of human intervention through fishing, marine organisms would be affected by the short time scale (weather) and long time scale (climate) variability in the ocean. Life histories of these organisms (how and where they grow and mature) have developed in response to the environmental conditions.

Basic fisheries biology

Benthic marine animals (e.g., lobster, crab, geoduck, and scallop) spend most or all of their adult lives on the bottom. Demersal species (e.g., cod and sole) spend most of theirs near or on the bottom. Pelagic species (e.g., salmon, herring, and capelin) spend most of their adult lives within the upper 100 m of the ocean. Many, if not most, of these species are very fertile. A single mature female can produce hundreds, thousands, and in some species millions, of eggs each year. The mortality of eggs, larvae, and juveniles must be extremely high, given that on average each spawning female needs only two eggs to hatch, grow, and survive to become adults to maintain the size of a population through replacement of the parents. Otherwise, the abundance of fish populations would explode. The fact that they normally do not suggests that *high fecundity is a life history strategy used by most fish and shellfish species to increase the probability that some offspring will survive in the face of harsh environmental conditions.*

If, on average, the number of offspring growing and surviving to a fishable size (i.e., the "recruits" to the fishery) is greater than two, then the population will increase in abundance over time; if less than two, then the population will decrease. The change from one year to the next in abundance of recruits is termed recruitment variation. In addition to the many and sometimes extreme effects of fishing, natural variation in recruitment causes natural variation in the number of fish in a population. Natural variations in growth cause variation in the size and weight of fish at a given age. For any given population, the stock biomass is the number of individuals at a given age, multiplied by their weight at that age, summed across all ages. The effects of fishing can be as great or greater than natural variation in determining the number of fish in a population and its biomass. Fishing removes individuals from the population or stock that would normally grow and spawn to provide future recruits. In extreme cases, the more fish that are removed, the more difficult it can become for the stock to remain self-sustainable.

Box 2.1. Density Dependence and the Abundance of Fish Populations

In most fish species the population of spawners is usually described in terms of spawning stock biomass (estimated in metric tonnes). In salmon species, it is referred to as escapement (estimated in numbers of spawning fish). Female spawners each produce a large number of eggs. Larger females usually produce more eggs than smaller females. Eggs that survive hatch into larvae (fry) and some of those survive to become juveniles and then adult recruits (usually described as the number of fish surviving to a given age or size where they are vulnerable to fishing gear). Survival rates are affected by natural processes and by fishing. There are two types of natural processes loosely referred to as either density-independent or density-dependent. Density-independent processes are those in which the proportion of fish surviving is not affected by fish density (i.e., abundance or biomass in a given area) but rather by some factor, such as fluctuating water temperature or salinity, that operates independently of fish density. In contrast, density-dependent processes are those in which the density of fish affects the proportion of fish that survive. For example, in Pacific salmon it has been frequently observed that the survival rate of juveniles before they venture to sea decreases as their abundance in freshwater increases. This results from many individuals competing for limited food or other limiting resources. In many fish species, body size at a given age also decreases with an increased density of fish, again mostly due to competition for food. Reduced temperature also can limit growth and thus have the same effect. Such decreases in size or weight of each fish at age leads to a tradeoff between the number of fish and the total biomass that is potentially harvestable (see Box 2.2). Density-dependent survival and (or) growth exist in a wide range of stocks and species.

Fish and the marine environment

The weather and climate of the atmosphere influence how and where humans live; similarly fish in the ocean are influenced by the properties of the water (temperature, salinity, currents, depth, light level, turbulence, etc.) in which they live. Animals in the oceans have evolved different strategies to survive in their habitat: continental shelf or deep ocean, coral reef or arctic seabed. These animals have evolved over tens of thousands of years in response to the ever-changing environmental conditions that were experienced by the ancestors of the fish that we catch today. We are gradually expanding our knowledge of fish in the marine environment, although our present understanding of the various interactions within and among the different species and their environment is still rather limited. Nonetheless we have begun to develop some ability to forecast the consequences of change, be it natural or man-made (anthropogenic).

The three coasts of Canada (Atlantic, Arctic, and Pacific) exhibit very different environmental conditions and have different marine fishery resources. Invertebrate species represent important fisheries on the Atlantic and Pacific coasts (e.g., lobster, crab, scallop, and shrimp in the Atlantic and crab, sea urchins, and geoduck [pronounced gooy-duck] in the Pacific). On the Atlantic coast, with its wide and shallow continental shelves, major fisheries rely on species that have evolved to exploit the shallow water environment and live near the bottom. Such demersal fish are called groundfish, with cod being a well-known example. There are nonetheless also important pelagic fish on the Atlantic coast that include, for example, herring and capelin. On the Pacific coast the continental shelf is less extensive than on the Atlantic and so demersal species make up a relatively small proportion of the fish community, which is dominated by pelagic or quasi-pelagic species (e.g., herring, salmon, Pacific hake, and rockfish). Here the different species of Pacific salmon predominate. Unlike the fish species covered in this report that spend their entire lives in the ocean, salmon spawn in freshwater. Their offspring migrate to the open ocean as juveniles (fry or smolts, depending on the species) and return as adults to their native river to spawn and complete their life cycle. The prevalence of pelagic species on the Pacific coast relative to the Atlantic may be due to the intense transport and cycling of food and energy through zooplankton in the pelagic environment rather than the settling of food to the benthos where it supplies bottom-feeding fishes and invertebrates.

Cod in the Atlantic

In the Northwest Atlantic there are three species of cod: Atlantic cod (*Gadus morhua), Arctic cod (Boreogadus saida*), and Greenland cod (*Gadus ogac*). The east coast cod fishery is focused on Atlantic cod, which are found throughout the North Atlantic from Georges Bank to northern Labrador in the western Atlantic, Greenland, and Iceland; the Barents Sea, Norway, and the Irish and North seas; and the Baltic (Fig. 2.1). These stocks are genetically different (e.g., Pogson et al. 1995) although there are observations that indicate some fish do occasionally migrate across the Atlantic (Templeman 1979; Taggart et al. 1995).

Cod are relatively long lived (20+ years in some places) and can grow to a metre or more in length and weigh as much as 50 kg. As adults they spend most of their time near the bottom feeding on shrimp and other benthic organisms and forage fish such as capelin and herring. In the Newfoundland region (at least until recently), the majority of adults spawn in late winter and spring on banks and along the edge of the shelf at depths up to 300 m (Hutchings et al. 1993; Rose 1993). Each mature female releases hundreds of thousands and even millions of eggs, which are then fertilized; only a very small proportion (e.g., a few 10 000ths of a percent) will survive to become spawning adults. The eggs, about a millimetre in diameter, rise to the surface layers of the ocean and hatch into larvae after several days or weeks, depending upon the water temperature.

By autumn, a few months after hatching, the larvae have grown to become pelagic juveniles several centimetres in length and they begin to settle to the bottom to begin their demersal life. The settled juveniles do not join the spawning adults in their migrations for

Fig. 2.1. Map showing the distribution of cod in the North Atlantic and the average annual reported landings (thousands of tonnes) during the 1965–1990 period for (a) Georges Bank, Gulf of Maine, Bay of Fundy, Scotian Shelf, Gulf of St. Lawrence, and southern Newfoundland, (b) the Grand Bank, Flemish Cap, and NE Newfoundland Shelf, (c) West and East Greenland, (d) Faroe Islands, (e) the North Sea, Kattegat, and Skagerrat, (f) Icelands and the NE Arctic, and (g) the Baltic and other regions. Landings data are derived from Garrod and Schumacher (1994).

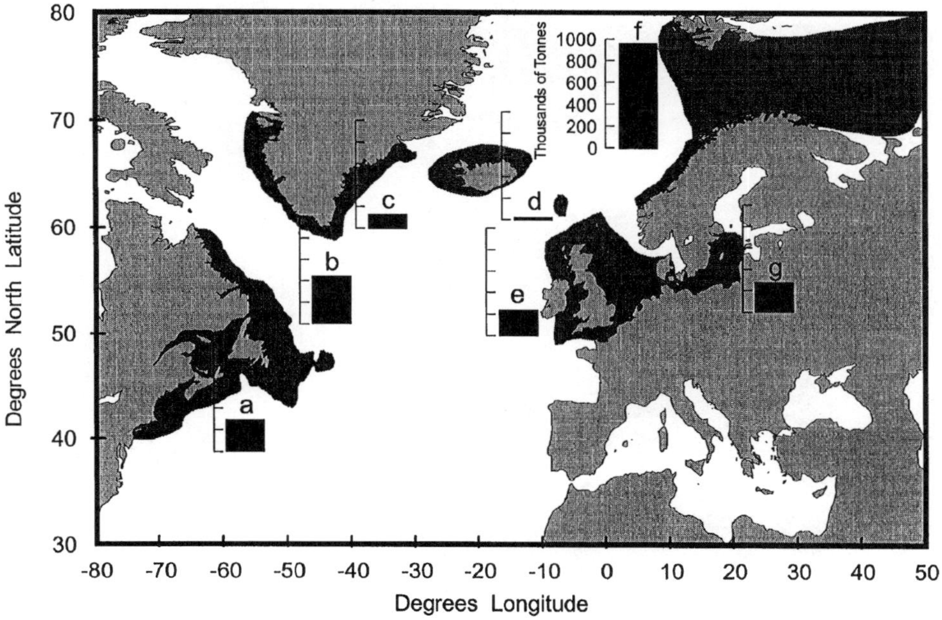

several years (Dalley and Anderson 1997; Taggart et al. 1994). The age at which a cod is ready to spawn depends upon its size, which is a function of its growth rate, which is, in turn, a function of the temperature the cod experiences while growing (Brander 1994, 1995). Cod in the Newfoundland and Gulf of St. Lawrence regions are some of the slowest growing cod in the world because of the cold waters found there. They take several years longer to reach spawning age than do cod that live on Georges Bank or in the Celtic Sea (Fig. 2.2). This simply means that the reproductive capacity and potential population growth rate for cod in these cold regions are much lower than cod elsewhere and thus they cannot be fished sustainably at rates as high as elsewhere (see Hutchings et al. 1997).

Cod in some regions (e.g., the Labrador Shelf when they were there and the Barents Sea) exhibit extensive seasonal migrations, yet in others they show little migration at all. On the Newfoundland Shelf, the majority of adult northern cod migrated inshore from the shelf break in the spring and in the late fall migrated back (Colbourne et al. 1997; Taggart 1997). There are so few northern cod left now that we cannot tell what has become of this migration pattern. Barents Sea cod show a similar offshore–inshore migration pattern except they spawn inshore as opposed to offshore.

On the Atlantic coast, seasonal cycles primarily determine oceanic conditions such as temperature and salinity. Circulation is somewhat seasonally dependent but the general patterns do not change much from summer to winter (Fig. 2.3). Winter cooling causes the surface waters to mix downward, sometimes all the way to the bottom, and in many locations

Fig. 2.2. The different lengths (cm) that age-4 cod achieve in different stocks in the North Atlantic. Cod living at colder temperatures (°C) take longer to grow to a given size than do cod at warmer temperatures. Data are derived from Brander (1995).

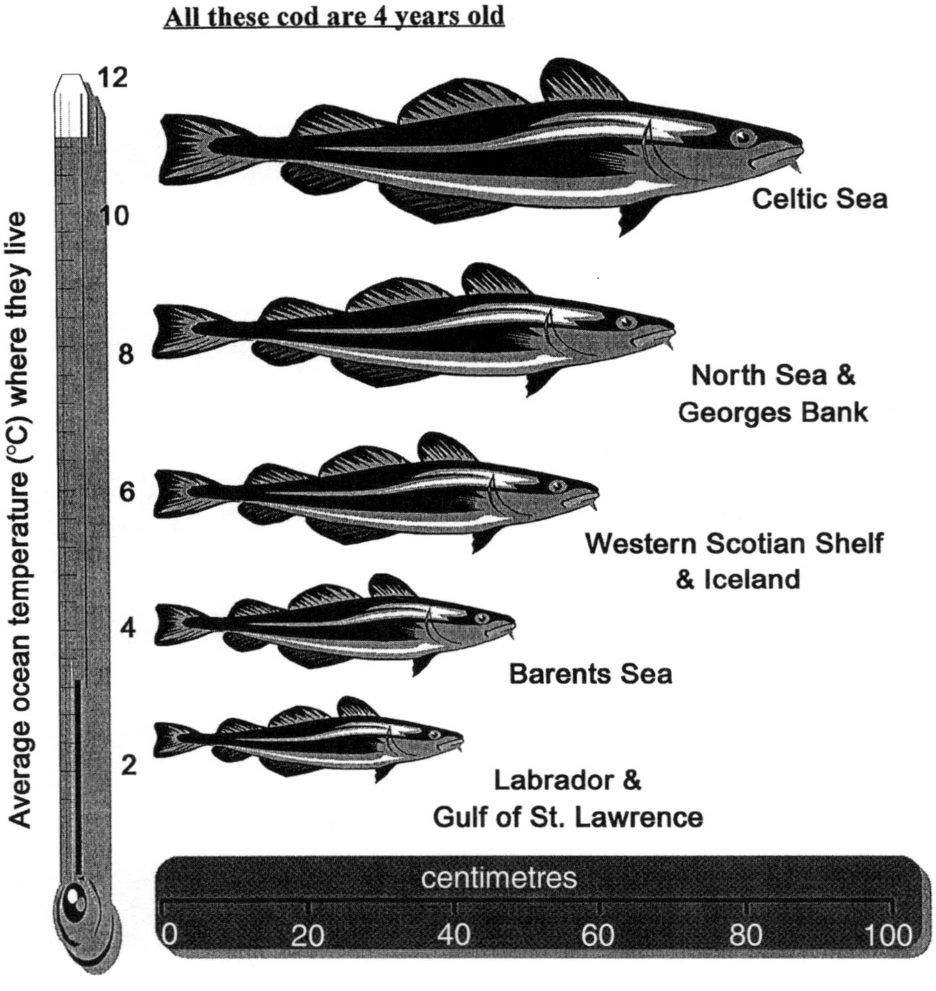

(e.g., off Labrador and in the Gulf of St. Lawrence) this leads to the formation of sea-ice. The southward transport of the ice by the currents and subsequent melting in early spring is a determining factor for surface water properties over much of the region. Deep waters are influenced by these processes as well as by exchanges across the shelf with the deep ocean.

The time of spawning of cod and their migration is generally related to the seasonal cycles in oceanic conditions (Myers et al. 1993*a*). Cod spawn so their eggs and larvae will be able to mature and feed in favorable spring/summer conditions when temperatures are increasing and food is available. The spawning cycle has evolved so that, on average, the chances for success are maximized for normal environmental conditions.

Fig. 2.3. Surface circulation in the Northwest Atlantic. The thickness of the lines is a rough indication of the strength of the currents.

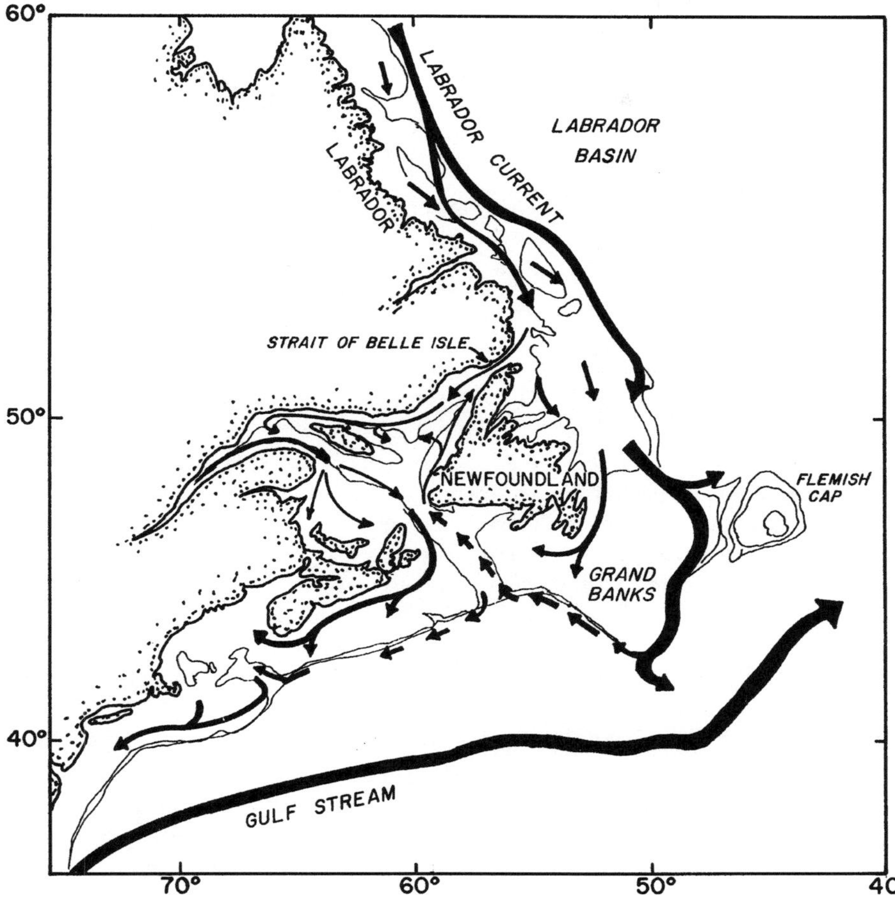

Salmon in the Pacific

There are six major species of salmon in the northeast Pacific: sockeye (*Oncorhynchus nerka*), pink (*Oncorhynchus gorbuscha*), chum (*Oncorhynchus keta*), chinook (*Oncorhynchus tshawytscha),* coho *(Oncorhynchus kisutch*), and steelhead (*Oncorhynchus mykis*). These species are anadromous (except for kokanee, the land-locked version of sockeye); that is, they spawn in freshwater, but at some life stage they migrate to the ocean, where most of their growth occurs (Fig. 2.4). Once mature, Pacific salmon return to their natal stream to spawn on gravel beds and then they die. This life history has resulted in the development of many genetically and reproductively isolated populations, often referred to as stocks, within each species.

There is a variety of life histories in different Pacific salmon species (Groot and Margolis 1991). Although all species require freshwater for spawning and early development, they differ in the time spent growing in freshwater. Pink and chum salmon migrate

Fig. 2.4. Map of the Northwest Pacific showing the location of some key rivers.

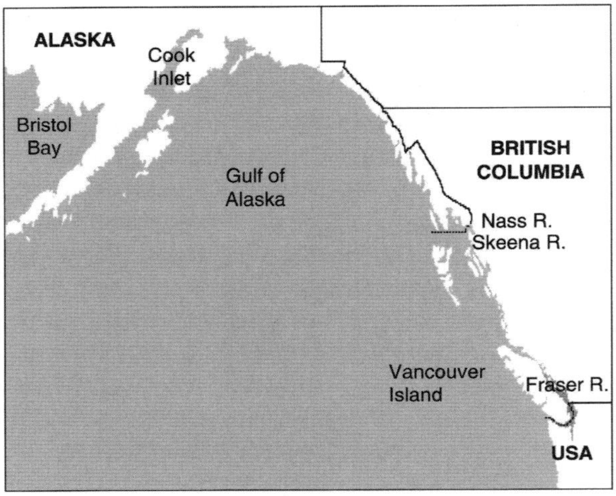

downstream to the ocean within a few weeks of emerging from river beds as fry. Chinook, sockeye, and coho normally spend one or more years in freshwater before migrating as smolts to the ocean. The amount of time spent in the ocean also varies considerably, ranging from 1.5 years for most pink and coho salmon to as long as 2–4 years for chum, sockeye, and chinook. When the salmon return to their natal streams they are between 2 and 7 years old, depending on the stock. The spawning migrations can cover distances as large as 3000 km for some stocks and may take Canadian salmon across Alaskan, or continental US boundaries, and into international waters. The wide variability of life history strategies in Pacific salmon has resulted from the unique adaptations of various stocks that allow them to take advantage of a diversity of freshwater and marine environments.

Salmon gain most of their size and weight while in the ocean. They feed along the British Columbia (BC) coast and in the Gulf of Alaska on various invertebrates (e.g., zooplankton and squid) and sometimes on small fish. Mortality rates for salmon cohorts (groups born in the same year) are moderately high in the ocean. For BC salmon the average proportion of wild fish that migrates seaward and then returns as adults toward their natal streams is typically less than 10% for sockeye and coho salmon and less than 3% for the other species (Bradford 1995). While these values reflect the average survival rates in the ocean, they are quite variable from year to year, as are the survival rates of the juvenile salmon leaving their rivers. Such variations are in part due to interannual variation in size at the time of seaward migration and in part due to variations in the physical and biological conditions that the salmon encounter in the ocean.

Seasonality plays a different role on the west coast, both in the physical environment and in its influence on the fish. The dominant effect of the seasonal cycle is the spring and autumn reversal of the prevailing winds and the consequent changes in current patterns and the alteration of the upwelling and downwelling cycle along the west coast of Vancouver Island and in Hecate Strait (Fig. 2.5). Wind-driven upwelling of deep, nutrient-rich water in summer gives rise to plankton blooms. This atmospherically driven seasonality in surface water properties and plankton production is qualitatively different from the classical model of warm stratified surface waters preceded and followed by plankton blooms in the spring and autumn. Non-salmonid fishes have adapted to this cycle by spawning before the upwelling begins and in places where the loss of progeny from the continental shelf (also a

Fig. 2.5. Surface circulation in the summer and winter for the Northeast Pacific.

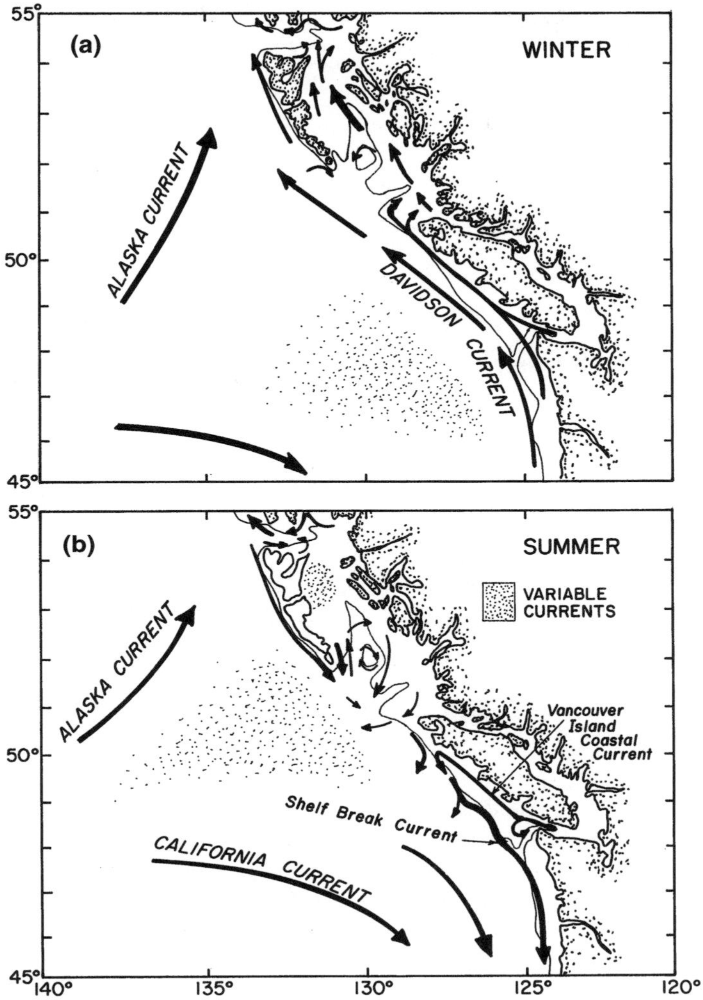

consequence of upwelling) is reduced. They have also adapted by timing their migrations along the shelf to those periods of maximum upwelling and therefore maximum plankton production. Juvenile salmon migrations during the late spring and early summer are a good example of adaptation.

Interactions and connections

In addition to the life history differences outlined above, there are other fundamental differences in how these two different species, cod and salmon, interact with their environment. As eggs and fry, salmon begin their lives in rivers or streams and lakes. Cod are completely marine and begin life as eggs and then yolk-sac larvae in the near-surface waters of the continental shelf where they are transported by ocean currents and may find themselves in environments that either enhance or reduce their growth and survival. Another fundamental difference is that cod larvae eat zooplankton while as adults they eat other fish

Fig. 2.6. Time series of survey-based assessment (1980 and 1992) estimates of age-4 northern cod recruits (solid line, open symbols) and modeled estimates of age-4 recruitment (dashed line, solid symbols) based on variation in the summer surface salinity off St. John's. After Myers et al. (1993*b*).

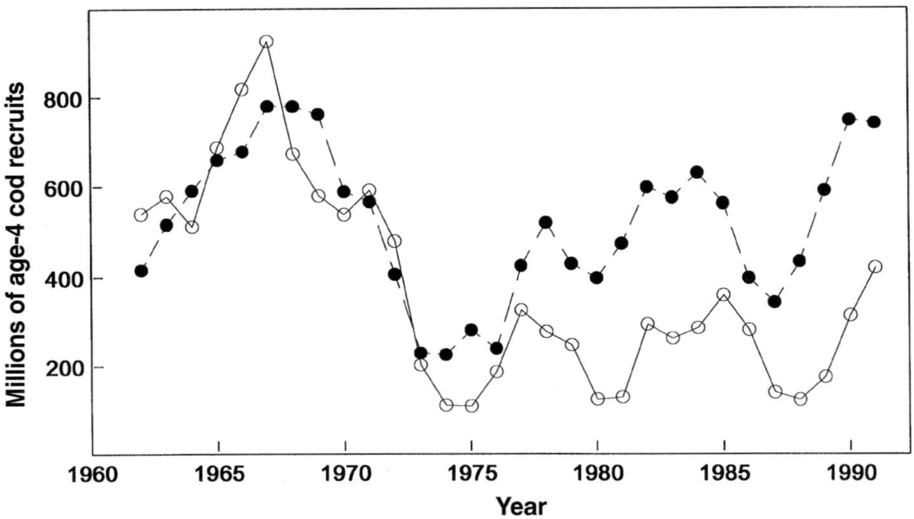

and benthic invertebrates. Salmon feed on invertebrates throughout their lives, although as adults some also eat fish.

In spite of the obvious connections between fish and their environment, attempts to quantitatively link environmental variation with variation in recruitment have, in general, met with limited success. For cod, there are hints of relationships with temperature (deYoung and Rose 1993) and some evidence that salinity may be important (Fig. 2.6). For salmon, success in establishing the influence of the environment on survival rate at sea has been equally limited. However, the coastal return migration route for sockeye salmon from southern British Columbia has been shown to be affected by sea surface temperature at the time of return (Mysak 1986).

The oceanography of the Pacific, Arctic, and Atlantic oceans is, of course, connected to the rest of the world and the strength and extent of global connections is critical to global change studies. In the Pacific the dominant large-scale feature of interest is ENSO, the El Niño–Southern Oscillation phenomenon (Philander 1990). At the beginning of an ENSO event, the effects of which can last for several years, easterly winds in the equatorial Pacific weaken. This leads to the formation of a large pool of warm water along the equator the size of North America, or greater, that may be several degrees warmer than is normal. This warm water pool contains a tremendous amount of energy, both because of the size of the temperature anomaly, roughly 5°C in 1997, and the large heat capacity of water. Its global impact is exerted through direct heating of the atmosphere by the ocean and the consequent dynamical response of the atmosphere, so-called atmospheric tele-connections. ENSO is the primary source of year-to-year and decade-to-decade variability in the North Pacific (Fig. 2.7) influencing both the atmosphere and the ocean. Its effects are also felt much farther afield, indeed, there is evidence that ENSO influences the frequency of occurrence of

Fig. 2.7. Long-term series of annual anomalies of the mean surface temperature measured at Amphitrite Point on the west coast of Vancouver Island. The large triangles at top indicate strong ENSO events; the smaller triangles indicate weak ENSO events. Warm anomaly peaks are clearly associated with the strong ENSO events.

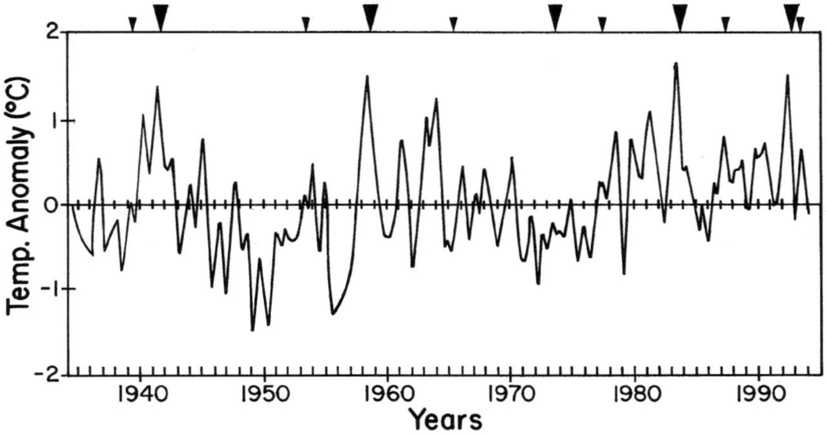

Atlantic hurricanes.

In the Atlantic most of the interannual atmospheric variation is associated with the North Atlantic Oscillation (NAO), which has been correlated with observed oceanographic variations (Dickson and Brander 1993). Atmospheric circulation over the North Atlantic is typically dominated by a low pressure centre near Iceland and a high pressure centre near the Azores. Thus, a measure of the NAO is the pressure difference between these two locations. When positive, the pressure difference generates increased winds from the northwest that bring cold arctic air southward and this leads to negative temperature anomalies (colder than normal) in the northwestern Atlantic. The highly positive NAO index in the early 1990s is associated with the colder than normal conditions observed in the Northwestern Atlantic (Fig. 2.8). In general, these colder than normal periods in the western Atlantic are associated with warmer than normal periods in the eastern Atlantic.

Fish stocks and their status

Estimates of the number of individual fish in a population or stock and the size and age distributions of those individuals are all essential for assessing its state, or health, in terms of numbers and total biomass. There are other less direct measures that can be made, but number, age, and size (meaning length or weight at a given age) are the fundamental measures. The number and size at age of individuals and their respective rates of change can be used to estimate not only how much biomass exists in a population, but also how well a population can sustain itself through reproduction in the face of fishing pressure and pressure arising from natural or man-made environmental changes. Size plays a key role because fish size, more than age, is the primary determinate of maturation and fecundity. In general, populations are more likely to be at risk of serious depletion when they drop below some level of abundance. If the decline in number is paired with a decline in the size or mean age of individuals, then it becomes even more difficult for the population to persist. However, estimates of numbers and size and age are exactly that — estimates — measures

Fig. 2.8. (a) North Atlantic Oscillation (NAO) index, (b) winter ice cover, (c) average water temperature off St. John's, and (d) average salinity off St. John's for the past four decades.

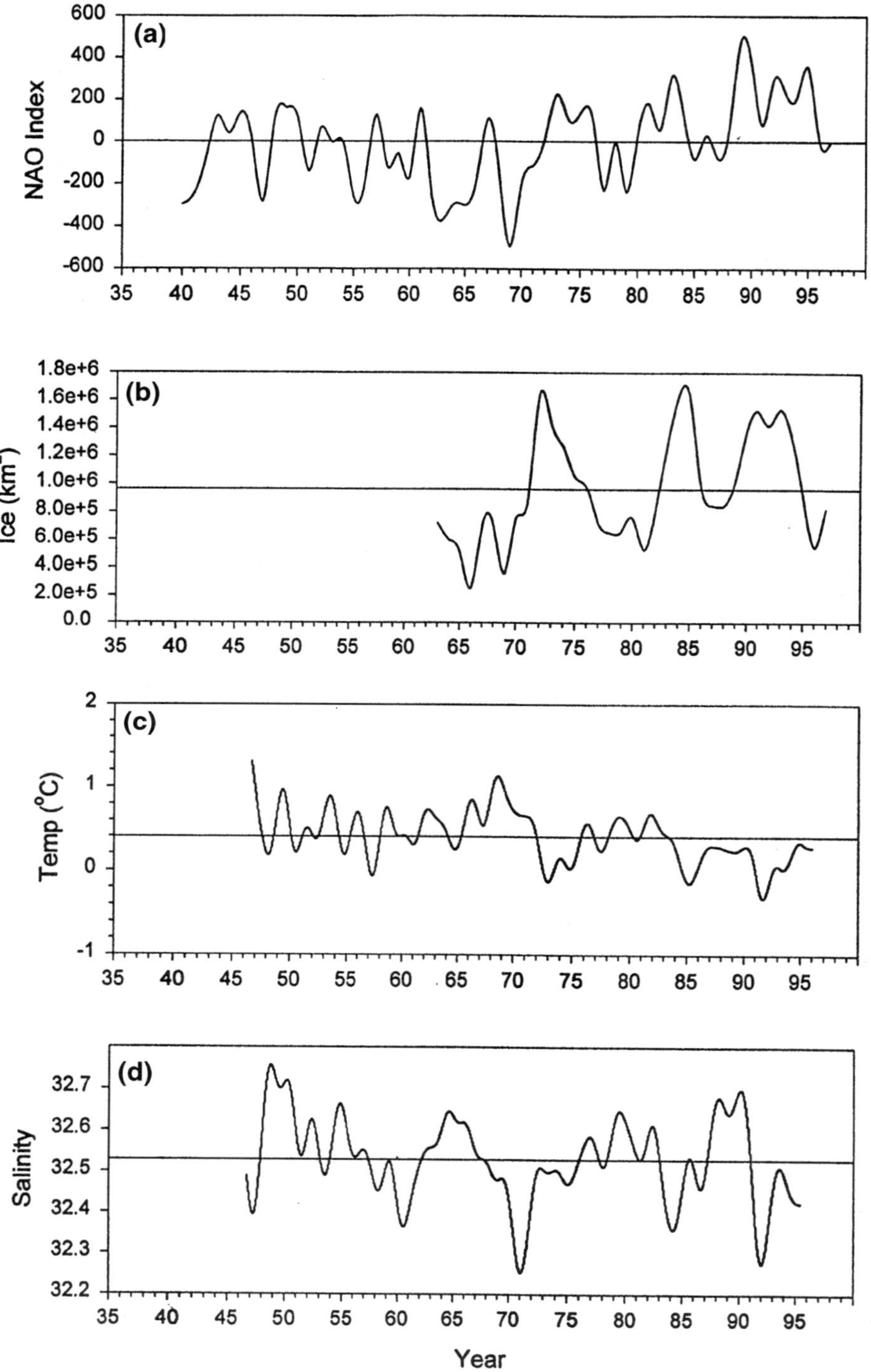

that have uncertainty associated with them and for fish populations there is generally a high degree of uncertainty.

Box 2.2. Sustainable Harvests

As noted in Box 2.1, growth and survival rates in fish populations are often density dependent in a way that tends to compensate somewhat for changes in abundance. For instance, reduced abundance of spawners often increases survival rates and growth rates (creating larger females that produce more eggs, thereby tending to increase abundance). This compensation creates the potential to harvest fish populations sustainably if the harvest rates are sufficiently controlled. For instance, in sockeye salmon, the number of adults (recruits) that is harvestable annually is a combination of two factors, the number of spawners and the survival rate of their eggs that will eventually become recruits. If the proportional survival rate is density-dependent in a way that survival rate decreases with the increased abundance of spawners (see Box 2.1 and Fig. 2.9a), then this produces a dome-shaped relationship between the number of recruits and number of spawners (Fig. 2.9b). Thus, some intermediate number of spawners will generate the greatest sustainable yield (i.e., number of recruits in excess of the number of spawners required to replace the spawning population through time). The number of spawners that gives rise to the "maximum sustainable yield" is one example of a "target escapement" that management agencies sometimes aim to achieve (see Box 2.9). However, note in Fig. 2.9b that many different levels of annual harvests or yields are sustainable over the long term. A moderate level of sustainable harvest, less than the maximum, is possible if the spawning population is held at a level above the one that would give rise to the maximum sustainable yield. This situation might be desirable in cases where there are management objectives such as avoiding severe stock depletion due to unusually low survival rates. Also note that this example of sustainable yields for a sockeye salmon population has a direct analogy with other species of fish, whereby sustainable harvests are defined in terms of harvest of "surplus production," or the excess of the annual increase in biomass (due to reproduction and to body growth) over the annual decrease (due to natural mortality). More details are given in Chapter 3.

Sustainable harvest strategies are in some cases based on "yield per recruit" analyses. These analyses are designed to estimate the total biomass at each age of a given cohort, or year-class (a group of fish that were born in a certain year; Fig. 2.10). Marine fish species such as groundfish and herring are often managed using a constant harvest rate policy (see Box 2.3). Then the management goal is to take some target proportion of the fish, or of the fishable biomass, at an intermediate size and age, neither too young, when there are many small fish, nor too old, when there are too few large fish. This procedure attempts to avoid what fisheries managers refer to as "growth overfishing," which is where a stock is fished too intensively to permit many fish to reach a large body size and hence where the biomass available is reduced from what it could be if more fish were allowed to grow larger before being harvested. However, many analyses of fisheries around the world (starting with Cushing 1973 and ending most recently with Hutchings and Myers 1994) have shown that too much focus on the yield per recruit approach has led to the implicit, and often incorrect, assumption that "recruitment overfishing" as defined above could not occur. Nevertheless, in some stocks where the stock biomass was reduced too far, recruitment overfishing did occur and severely reduced stock biomass, even when the yield per recruit approach was being followed.

Fig. 2.9. (a) A typical relationship for Pacific salmon between the number of adult recruits produced per spawner and abundance of spawners, illustrating the effect of density-dependent survival rate on reduced reproductive success at high spawner abundance. (b) Total number of adult recruits produced by a given spawning population, as a result of the relationship in part (a). The dashed "replacement" line indicates the number of recruits required to just replace the spawning population. Different sustainable yields (c) are calculated by the vertical difference between the recruitment curve in (b) and the "replacement" line, thereby indicating the excess of recruits above the number of spawners required to replace the population.

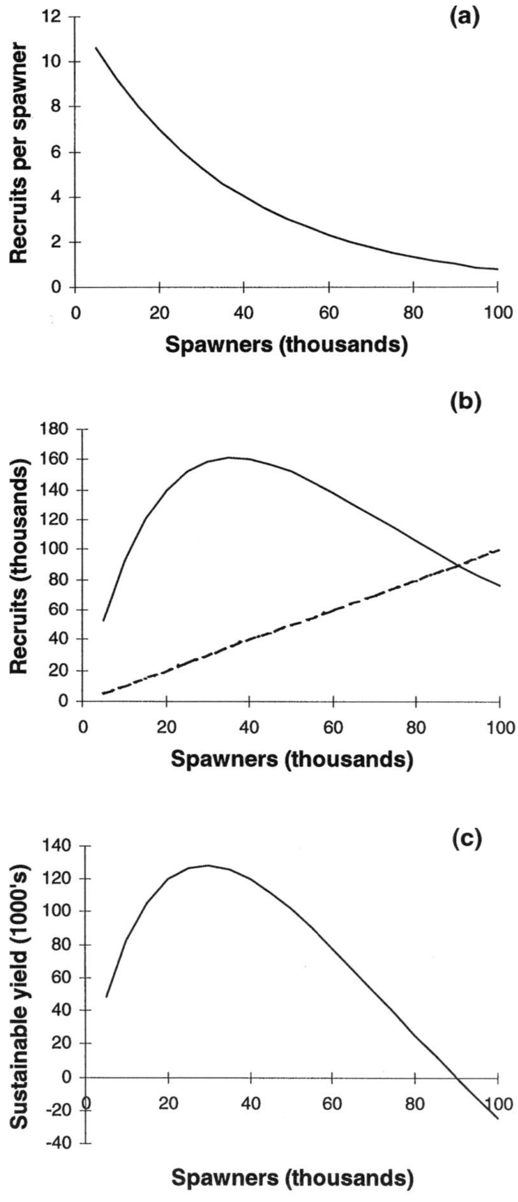

Fig. 2.10. Typical fate of a cohort of marine fish (i.e., individuals that were all born in the same year). (a) Number of fish surviving in the cohort at each age, (b) the body size (kg) of each of those surviving fish at each age, and (c) the total biomass of the entire cohort at each age (equal to the number still surviving multiplied by their body size at that age).

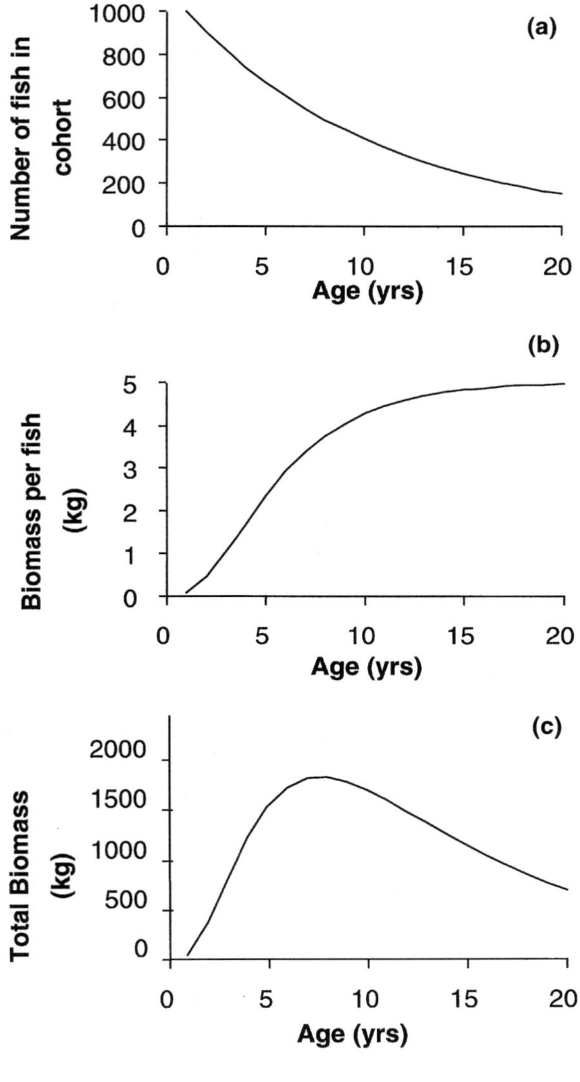

Assessing the status of fish stocks

In order to estimate the state or health (lengths, weights, condition, sex ratios, age distributions, number of new recruits, etc.) of a fish stock, one must monitor the stock by taking a variety of measurements, analyzing the data, and drawing inferences from the results, a process called "stock assessment." The many ways of achieving this are dependent on the nature of the species assessed and on the cost of getting the measurements. However, in order to apply almost any method, many assumptions must be made. The most important is that the geographic distribution or range of the stock is known and that the stock is resident in that area at the time it is assessed — simply defined as "availability." If one assumes that the fish are fully available to be assessed when in fact they are not (say, for example, it is not known that some of the population has migrated away), then the assessment will be wrong. This problem is obviously more important for migratory species (cod, salmon, herring, etc.) than it is for more sedentary species (e.g., crab).

How then should one assess a fish stock? The simplest and least expensive method is to use the catch or landings data from fishing vessels. However, catch or landings data on their own can be misleading in drawing conclusions about the state of a stock. To use such data several criteria must be met. The first is that the fishing effort applied must either be constant or known. However, effort is rarely constant and is difficult to measure primarily because of the expanding and diverse technology used in fisheries. Consider the catch resulting from the effort of six fishermen in one boat using the same gear and same methods from year to year. If the catch goes up or down from year to year, then it is reasonable to conclude that the size of the population is going up, or down, in the same relative manner. The availability assumption outlined above must also be met to reach this conclusion. However, if the fishermen learn more about where and when to find the fish from year to year and they use increasingly sophisticated technology (a faster boat, an improved echo-sounder, satellite navigation, etc.) to seek out and catch the fish more efficiently from year to year, then it is entirely possible for catch rates (fish caught per boat per day) to increase from year to year while the actual number of fish in the stock is declining. Thus, without a correction for the "real" increase in effort provided by learning and new technology, the wrong conclusion easily can be reached that the stock is increasing. It is possible to make a correction for one boat, however, it is more difficult to do so from year to year for a fleet of large and small vessels using a wide array of old and new technology and for fishermen with a range of skill and experience levels.

A second, and equally important criterion needed for using catch data is that all catches and landings are reported accurately (what, how much, when, and where caught), *including* those fish that may be discarded or "dumped" at sea. Even fish of the right species are dumped at sea if they are of the wrong size in a market sense (a process known as high-grading). Harvest fisheries in Canada and throughout the world are frequently regulated in a manner that provides incentives for misreporting landings and discarding. Numerous fisheries are also size-limited and so most, if not all, of the small fish that would recruit to the future fishery and would provide future spawning potential, are killed in the process of being caught and are discarded because they cannot be legally landed. If unreported, these fish cannot be assessed as having been removed from the stock, thereby leading to overestimates of the remaining fish.

It is clear from the above that if the criteria related to fishing effort and the reporting of catch are not met, then it becomes possible to maintain very high catch rates over some period or even increasing catch rates, independently of the stock abundance, even if it is declining. It is because of these various aspects of the harvest fishery that *using catch data alone to infer the status of a stock can be dangerously misleading and, at best, uncertain.*

An alternative method to assess year-to-year variations in fish abundance and (or) biomass is to survey or "sample" the population in a manner that is similar to a human population sampling survey (e.g., a political poll). The survey techniques vary from fishery to fishery and for different species. In a typical groundfish survey, like that employed for cod, fishing effort in the form of a research vessel (RV) survey is kept constant from year to year by using the same vessel and fishing gear and sampling in a manner that permits coverage of the entire geographic distribution normally expected for the stock. It is analogous to the "effort applied by six fishermen in one boat using the same gear and the same methods from year to year" considered above. Of course, using the RV survey successfully also assumes that the entire population (or at least the same proportion from year to year) is available in the region when it is being assessed. In general, the entire region is divided into a number of blocks or "strata" which reflect various fish habitats (e.g., depth) and then samples are taken (i.e., fished) systematically but at randomly chosen locations within each of these strata. More samples and more strata should provide a better estimate, however, at greater cost. The "sample" catch, which is assumed to be representative of the entire population, is then counted and individual fish are examined for length, weight, sex, maturity, age, etc. The survey results are then used to estimate the stock status, as well as the degree of uncertainty in the inferences drawn. The most frequently used estimates of the state of a stock are its total biomass (total weight in tonnes), the spawning biomass (total weight of fish large enough to spawn), or the biomass of certain age or size-classes. Advanced echo-sounder technology can also be used in a similar manner, though with different and more complex uncertainties, to arrive at similar estimates.

The RV survey approach, when used in concert with commercial catch data, is often used to "reconstruct" the abundance and age distribution of the fish population as it existed through time using methods known as cohort analysis, or virtual or sequential population analysis (VPA or SPA). Such analyses depend of course on reliable input data and the poorer the data, the worse the reconstruction. Furthermore, the more recent in time the estimate, the more uncertain it is. Forecasts of abundance into the future are particularly uncertain. In general, however, this method is far more reliable than using catch statistics alone.

The assessment of salmon is somewhat different and unique, in terms of typical marine stock assessments. It is based in part on estimating, through the use of counting fences, sonar, or tag-and-recapture studies, etc., the number of reproductive adults when they return to their native rivers to spawn. This is most frequently achieved through estimating "escapement" — the number of spawning salmon that escape the fishery to reach the spawning grounds each year. When the escapement is added to the number of fish from the population that did not "escape" and were caught and landed, then the total number of recruits for that year's "run" can be estimated. This kind of assessment also requires certain critical assumptions regarding the fishery-independent estimates (such as those derived from tag-and-recapture methods) that are analogous to the RV survey, as well as assumptions regarding the amount of commercial catch and its composition of different stocks. As in groundfish assessments, these assumptions are not always met and, thus, lead to uncertainty in the estimates of salmon stock status.

Descriptions of the general state of the major Canadian fish stocks, presented below, are to a large extent based on the different techniques described above.

The fisheries of Atlantic Canada

"The fish production systems of Atlantic Canada are in a state of crisis, in the social, political, economic, and biological senses" (Kerr and Ryder 1997, p. 1190). The state of the "crisis" from a biological perspective is that well over half of the major groundfish species and stocks in Atlantic Canada can be classified as unhealthy. This is consistent with the Food and Agriculture Organization (1995*a*) summary that states most of the important groundfish species throughout the NW Atlantic (Canada and the USA) are considered either fully exploited or overexploited. This includes species such as cod, redfish, pollock, and menhaden. Hake, haddock, and most flatfishes are considered fully exploited or depleted. Pelagic species such as capelin and herring are considered recovering and only mackerel was considered underexploited. All crustaceans (crab, lobster, shrimp, and prawn) are considered fully exploited, along with the major shellfish species (scallops and clams). The FAO points to ongoing high-grading and discarding as major obstacles that continue to distort estimates of total removals and therefore the population estimates. Stocks of demersal species in the northeast USA are also in a poor state, and even though the groundfish index in 1992 was at it lowest for the last 30 years, fishing effort has been increasing. Species diversity is changing and species replacement may have occurred as inferred through apparent increases in the abundance of elasmobranchs (sharks and rays). Although many of the groundfish species on Georges Bank are exploited by both the USA and Canada, there is no formal cooperation in management. On the positive side, Georges Bank herring, which have been commercially extinct (i.e., so low in abundance that it is not economical to harvest them) since the late 1960s, are now showing signs of recovery.

A focus on Atlantic cod

There are more than 17 recognized cod stocks in the North Atlantic. Subsequent to a highly productive period of the 1950s and 1960s, virtually all of them have suffered a protracted decline in abundance. During this same period there has been an Atlantic-wide increase in fishing pressure (Garrod and Schumacher 1994; Daan 1994). On average about 70% of an Atlantic cod year-class is being harvested annually (Daan 1994). Although comparisons of catch trends (while acknowledging the uncertainties associated with catch data) in some stocks provide some evidence of climactic effects, overexploitation remains the simplest and most difficult-to-refute explanation of recent stock declines. This is clearly illustrated in Fig. 2.11 where increasing fishing mortalities (F), an index of the removal rate

Fig. 2.11. Time series of the variation relative to the average in exploitable biomass (solid line, open symbols) and in the fishing mortality (dashed line, solid symbols) in northern cod during the period 1963–1984. Data are derived from Garrod and Schumacher (1994).

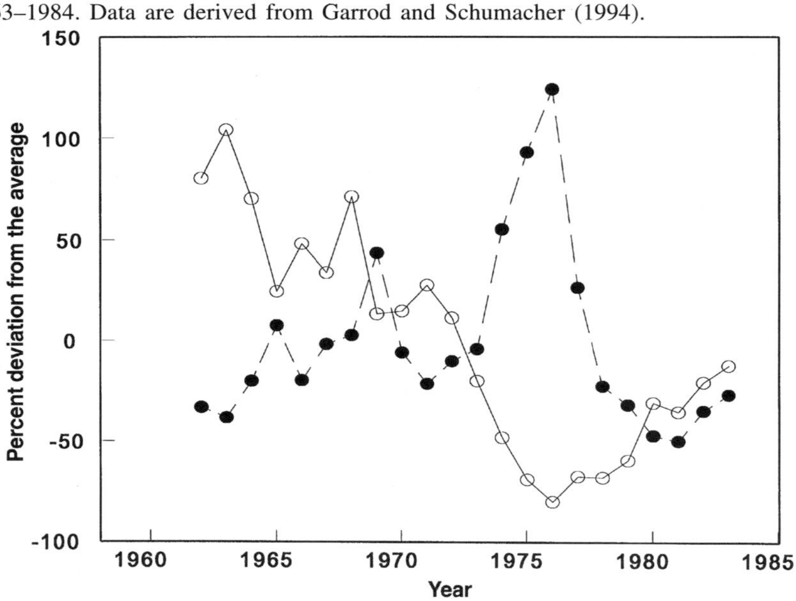

from the population due to fishing, are associated with decreasing population size estimates. Environmental variations have the *potential* to influence cod production in the North Atlantic (Garrod and Schumacher 1994). However, analysis of North Atlantic recruitment data (Myers et al. 1995) reveals no large-scale pattern of synchrony, perhaps suggesting that there is no common environmental factor regulating recruitment across the entire Atlantic. Fishing Atlantic cod has changed the age and size structure of populations to younger and smaller fish, on average, making the populations less productive and more vulnerable to unfavorable environmental conditions (details in Chapter 3). Daan (1994) points to a variety of physical processes among different regions and stocks that are differentially important to growth and survival of cod eggs and larvae (e.g., variation in retention on Georges Bank, variation in circulation and transport in the North Sea, and variation in the exchange of oxygen laden water masses into the Baltic). Thus, climatic change or year-to-year environmental variation may influence different stocks in widely different ways.

Box 2.3. Risks Associated with Environmental Changes

Under normal environmental conditions, the number of fish produced per tonne of spawning biomass (or per thousand spawners, for instance) would be at some level for a given spawning biomass. However, if some environmental change occurs that is "unfavorable," then the number of fish that survive to a given age is diminished along with recruitment and the number of fish harvested may have to be reduced to avoid overexploiting the population.

Cod in Canada

A recent examination of the Georges Bank and Gulf of Maine cod stocks (Serchuk et al. 1994) shows that record high fishing mortalities during the late 1980s and early 1990s resulted in marked reductions in the biomass of these stocks to record low levels. The cod on Georges Bank that are fished by Canadians are the fastest growing of all Canadian cod (see Fig. 2.2). Generally increasing fishing mortality rates in the last decade have been associated with decreasing trends in population size, length at age, and age at maturity. Female cod that used to mature at lengths of 50–55 cm and at an age of 2–3 years in the 1970s matured at lengths of 30–40 cm and at ages of 1–2 years in the late 1980s (Serchuk et al. 1994). Relative to older larger cod, small young cod produce fewer eggs, fewer of them are successfully fertilized, and their hatching success is lower (Trippel 1998). In 1996 and 1997, the Georges Bank cod stock was the only one of two cod stocks in Canada (the other being the western Scotian Shelf stock) subject to fishing, though at unprecedented low levels (DFO 1996*a*, Groundfish Management Plan).

According to Frank et al. (1994), the eastern Scotian Shelf cod stock biomass estimates are at near record low levels (Fig. 2.12) and the most recent recruitment estimates are the lowest on record (Fanning et al. 1996). This is a result in part of high fishing mortalities that have been well-above target levels (Frank et al. 1994). The environment has been shown to play an important role here. For example, freshwater discharge from the Gulf of St. Lawrence is thought to influence the recruitment dynamics of cod on the eastern Scotian Shelf (Frank et al. 1994) and recent analyses indicate that temperature and seal predation may be contributing to reduced recruitment (Fanning et al. 1996) in an already depressed stock. However, contemporary analyses suggest that the size of the spawning stock is currently

Fig. 2.12. The time series of standardized biomass estimates for 5 major cod stocks in eastern Canada: 2J3KL northern cod, 3NO southern Grand Bank cod, 3Pn-4RS northern Gulf of St. Lawrence cod, 4T southern Gulf of St. Lawrence cod, and 4VsW eastern Scotian Shelf cod. The data are derived from Davis et al. (1996), Fanning et al. (1996), Fréchet et al. (1994), Shelton et al. (1996), and Sinclair et al. (1996).

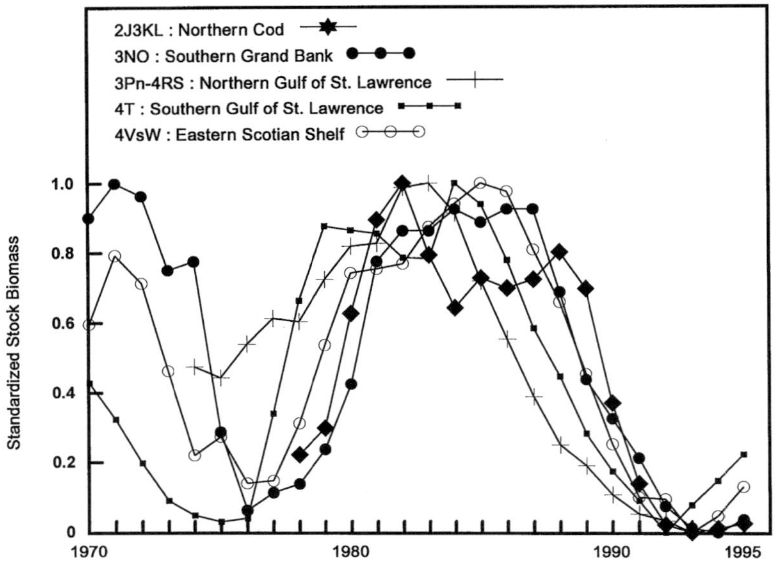

more important to future recruitment than are environmental factors such as freshwater outflow. There is additional evidence of altered reproduction patterns that are consistent with the loss of spawning components within the stock.

Cod in the Gulf of St. Lawrence are some of the slowest growing cod in the entire Atlantic (see Fig. 2.2) and size at age and recruitment in recent years have been below normal for this stock. Estimates of the stock biomass have been declining since the mid-1980s (Fig. 2.12) under relatively high fishing rates until 1993, and recruitment failure appears to be an important contributor to recent declines (Chouinard and Frechet 1994). Recruitment has been unable to compensate for removals by fishing and natural mortality. Although recent years are typified by lower than normal water temperatures and greater ice coverage in winter, there is no strong correlation between these environmental measures and the survival or recruitment indices.

The Newfoundland cod stocks, historically representing the most abundant of all cod fisheries in the NW Atlantic, have been closed over most of the Newfoundland and Labrador shelves since 1992 due to stock collapses (Hutchings and Myers 1994). Northern cod (the 2J3KL stock complex) and southern Grand Bank cod (3NO) spawner population estimates decreased to record minima in 1993 (Fig. 2.12) and have remained low since then. Limited fishing began in 3Ps, on the south coast of Newfoundland, with a quota of 10 000 t for cod (not northern cod) in 1997. There is also growing pressure to reopen fisheries, in spite of evidence that goundfish stocks remain in a depressed state over much of the shelf.

The decrease in abundance of northern cod was also associated with a collapse in the size and age structure of the stock (Taggart et al. 1994). Recruitment in northern cod relies heavily on spawner population size — the only NW Atlantic cod stock to show such a clear relationship (Hutchings and Myers 1994). Lengths at maturity have been generally declining since the early 1980s, a trend that is most pronounced in northerly regions (Morgan et al. 1994). The fishing mortalities on northern cod began to increase after 1980 and accelerated after 1987 to exceedingly high levels, particularly in the older large fish (Shelton and Morgan 1993). The combined effect of fishing mortality on the spawning population and the reduction in the length and age distributions of the stock had a direct effect on the amount of recruitment required to replace the spawner stock; i.e., from 1983 onward the stock was unable to sustain itself under the fishing pressure applied (Taggart et al. 1994). There has been much discussion of the role of seals, however to date, there is little evidence that seals played a significant role in the decline of the cod. Discussion is presently focused on the

Box 2.4. The Northern Cod Fishery

Development of the northern cod fishery over the past few centuries illustrates the effects of fishing technology on fishery systems. Hutchings and Myers (1995) note that annual harvests of northern cod remained at a low and relatively stable level of under 100 000 tonnes for two centuries from the early 1500s through to the late 1700s. A series of technological advances in the way of maneuverable dories and multi-hooked line trawls in the 1790s, side-hauled otter trawls in the 1890s, and various other trawling methods then followed. These changes, along with increases in the numbers of those fishing, helped to produce annual catches that varied around 200 000 tonnes over a period of roughly 150 years through to the mid-1900s (reaching 300 000 tonnes in the 1880s and 1910s). Based on the longevity of that fishery and despite improved efficiency (greater effort), each of these early harvesting strategies appear to have been sustainable, although there is evidence of problems in the mid-1800s (Cadigan 1995).

Around 1954, stern-hauled otter trawlers, notably the very large factory freezer trawlers, were introduced to the northern cod fishery. This significant increase in effort is considered "the single most important event in the five-hundred-year history of the northern cod fishery" (Hutchings and Myers 1995) and led to dramatic increases in annual catches. Indeed, this might be viewed as a regime shift in technology; a sudden change in gear composition of the fleet which in turn led to a rapid and imperfectly controlled increase in efficiency. The historically unprecedented spike in catches is clear (Hutchings and Myers 1995); reported catches reached a high of 810 000 tonnes in 1968, making northern cod, for a time, among the biggest fisheries in the world. Under such fishing, the stock could not sustain itself and it underwent its first collapse, in the early 1970s.

This first "crisis" led to the establishment of a fishing quota system in 1974 and the 200-mile exclusive economic zone (EEZ) 1977. It was optimistically thought that the Canadian fishery within the EEZ could manage a recovery and that the total catch could be increased to 450 000 tonnes by the end of the 1980s. Through the 1980s, however, a number of warning signs began to appear: declining weights and lengths at age, fewer older fish, changes in maturity, etc. Although the adopted fishing target levels (referred to as $F_{0.1}$ levels) and scientific recommendations suggested a reduction in the 1989 quota to 47% (125 000 tonnes) of what it was in 1988 (266 000 tonnes), the management decision was a reduction to only 88% (235 000 tonnes) of the 1988 quota (Shelton and Atkinson 1994). In fact, the quotas set in each of 1989, 1990, and 1991 were nearly twice that determined by the $F_{0.1}$ target (1.9, 1.6, and 1.9 fold higher each year respectively) (Shelton and Atkinson 1994). It is clear in hindsight that the actions taken were insufficient to halt the decline in the stock et al. 1997) and by the early 1990s it was recognized that the stock was inserious trouble. In July 1992 the northern cod stock was closed to fishing and several other stocks were subsequently closed. By 1998 the stock still showed no significantsigns of recovery and the prognosis for the future remains bleak.

What is the ultimate cause of the collapse and what could have been done to avoid it? There has been much discussion about the relative importance of overfishing and environmental conditions (e.g., anomalously cold water masses in the region (deYoung and Rose 1993; Hutchings and Myers 1994)). Overfishing of cod began in the 1950s and, with the exception of a short respite during the late 1970s and early 1980s, it continued up until the fishing closure in 1992. There is indirect evidence of stable (i.e., sustainable) catches for well over a century long before the 1950s during which period other extreme environmental events occurred, and yet there were no signs of a major collapse of cod like those observed recently. It would appear that given a sufficient stock size, particularly spawning stock (i.e., the capacity to reproduce and sustain itself), northern cod are well-adapted to withstand the vagaries of natural environmental fluctuations.

Overfishing was the primary cause of the collapse of northern cod (Hutchings and Myers 1994; Taggart et al. 1994; Myers et al. 1995; Hutchings 1996) although other factors may have played a role in the trajectory of the collapse. The collapse may very well have been avoided through reduced fishing effort in the late 1980s and early 1990s when the warning signs began to appear (Alverson 1987; Harris 1990), but uncertainty played a role here as well. Thus the response of government to various reports, scientific findings, and observations from fishermen have all been the subject of much debate — a clear reflection of the uncertainty.

In the end it will likely be shown that technological developments were not alone in steering the development and the collapse of the northern cod fishery, nor were scientific uncertainties, recognized or not. The fishery was a typical fishery system. Economic factors, consumer demand, size of markets, access to markets, capitalization of the fleets and processing plants all played a role, as did socio-political factors. The northern cod stock complex undoubtedly varies with environmental conditions and is likely affected by human impacts on the environment. However, it has been argued persuasively (Cadigan 1995; Hutchings and Myers 1995 and references therein) that change in the technological capabilities of the fishery was a major determinant of the dynamic changes seen in the fishery and its ultimate collapse. Furthermore, the response, or lack thereof, to changes in technology reflected fundamental attitudes toward exploitation and management of fishery systems that contributed greatly to the collapse (Charles 1995).

possible role that predation, primarily by seals, may be playing in limiting the recovery of the northern cod which has shown few positive signs since the imposition of the moratorium in 1992.

The fisheries of Pacific Canada

Relative to the Atlantic, particularly the NW Atlantic, Pacific stocks are in seemingly better shape, at least in general, for groundfish, pelagic species, and salmon (FAO 1995*a*). In the NE Pacific (Alaska and the West coast of Canada), the FAO as of 1992 considered all Pacific salmon, Pacific halibut, and Alaska pollock fully exploited, ocean perch depleted, and Pacific cod, hake, sablefish, and most flatfishes moderately exploited. Landings

Table 2.0. A qualitative summary of the status of the 12 major groundfish stocks in Pacific Canada as inferred from the Pacific Stock Assessment and Review Committee.

Species or species group	Status
Dogfish	Average to high
Sole (rock, English, Dover)	Average to high
Lingcod — offshore	Average
Pacific hake	Average
Sablefish	Average
Pacific cod	Low to average
Walleye pollock	Low to average
Rockfish (inshore)	Low to average
Rockfish (shelf slope)	Low to average
Rockfish (shelf)	Low to average
Lingcod — Strait of Georgia	Very low
Petrale sole	Very low

reported by the USA accounted for close to 90% of all landings from the NE Pacific in 1992. Of the 12 major Canadian groundfish stocks in the Pacific (Table 2.0), seven are considered to be have low to average abundance. Two are at very low levels and only 5 stocks (42%) are considered to be in average to better-than-average condition.

Increased US landings have occurred for sockeye and pink salmon but declines have occurred for landings of other species. Major threats to salmon stocks continue to be high harvest rates and anthropogenic degradation of spawning and nursery habitats. Recent disagreements between Canada and the USA over the allocation of catches continue to plague several salmon stocks. Groundfish catches have been generally declining following the increases experienced from the mid-1970s to the mid-1980s. None are considered critical with the exception of pollock in the central Bering Sea, which have experienced severe multinational overexploitation. Pelagic stocks (e.g., BC herring and Atka mackerel) show no clear trends.

A focus on Pacific salmon

Catches of BC stocks of Pacific salmon show several important features. Total tonnage of catch has remained relatively constant over the last 4 decades (1950s to 1980s) with a slight tendency to increase until 1994 (Fig. 2.13). Catches have remained generally low for chinook and coho and generally high and variable for sockeye and chum salmon, but pink catches have dropped dramatically since 1991. Chinook and coho catches have been trending downward over the last 20 years, in part reflecting losses of various stocks. In contrast, sockeye salmon catches have tended to increase over that period. However, as pointed out earlier in this chapter, assessing the status of a stock on the basis of catch alone can be uncertain or misleading. If the status of Pacific salmon stocks is assessed on the basis of

Fig. 2.13. Time trends in commercial landings (metric tonnes) of the five major Pacific salmon species. 1994 data are preliminary. Data from Canada Dept. of Fisheries and Oceans.

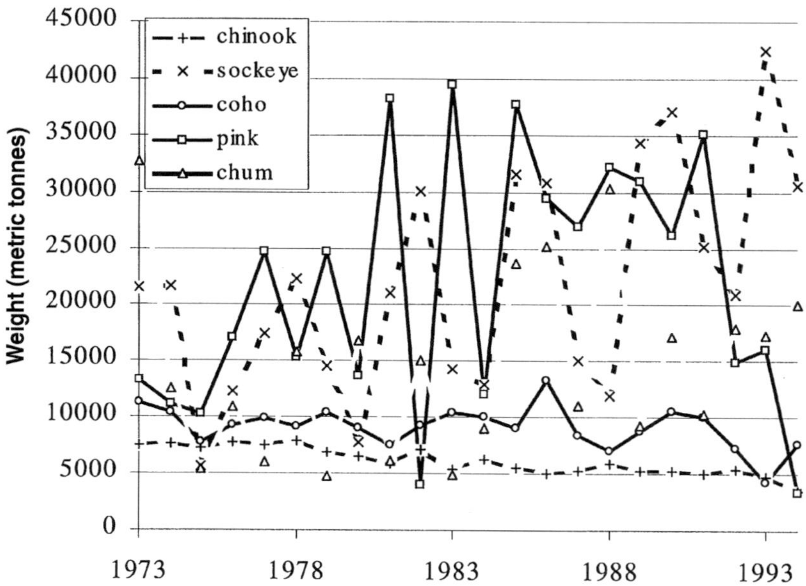

abundance rather than catches, then their status is not as favourable as that implied by Fig. 2.13, especially when we consider the losses of particular stocks that contribute to overall biological diversity.

Slaney et al. (1996) identified 9663 stocks of anadromous salmon and trout in British Columbia and the Yukon, many of which are relatively small populations of a few thousand fish or less. While 46% of the total stocks were categorized as unthreatened, more alarmingly, 43% of the stocks (4172) have an unknown status because of lack of data. Some (1.5%) of the stocks are extinct and 6.5% of them are at high risk of extinction. The Strait of Georgia, Johnstone Strait, and Southwest Vancouver Island regions have a higher proportion (14–17%) of high-risk stocks than most other regions. These average figures do not indicate other important features, for example, that a higher proportion of coho stocks is threatened than sockeye salmon, or that 16% of steelhead stocks are of special concern because they are regularly intercepted in net fisheries targeted on other species (such as the more lucrative sockeye salmon) (Slaney et al. 1996).

Although few data series exist for total abundance of wild chinook salmon, the abundance of spawning chinook salmon from the lower Strait of Georgia (excluding Fraser River stocks but including wild stocks and hatchery fish that stray and spawn in streams) has declined steadily (Fig. 2.14). Declines are more severe in mainland streams than in Vancouver Island streams. Although the catch of chinook salmon in the Strait of Georgia sport and troll fisheries was maintained at relatively high levels until the early 1980s, it subsequently experienced a monotonic decline to near record low levels by 1990.

Despite the incompleteness of the data, a similar situation is seen for wild Fraser River coho salmon (DFO 1996*c*). An index of abundance of this group of 187 stocks (escapement plus catch by commercial fisheries in Area 29 and Native in-river fisheries) has declined noticeably since the 1970s and dropped sharply in the early 1990s (Fig. 2.15). Note that this

Fig. 2.14. Time series of escapement for chinook salmon in the lower Strait of Georgia area, including fish produced by hatcheries but excluding Fraser River stocks. Escapement estimates from 155 streams as determined by DFO visual spawner estimates. Adapted from Levy (1996).

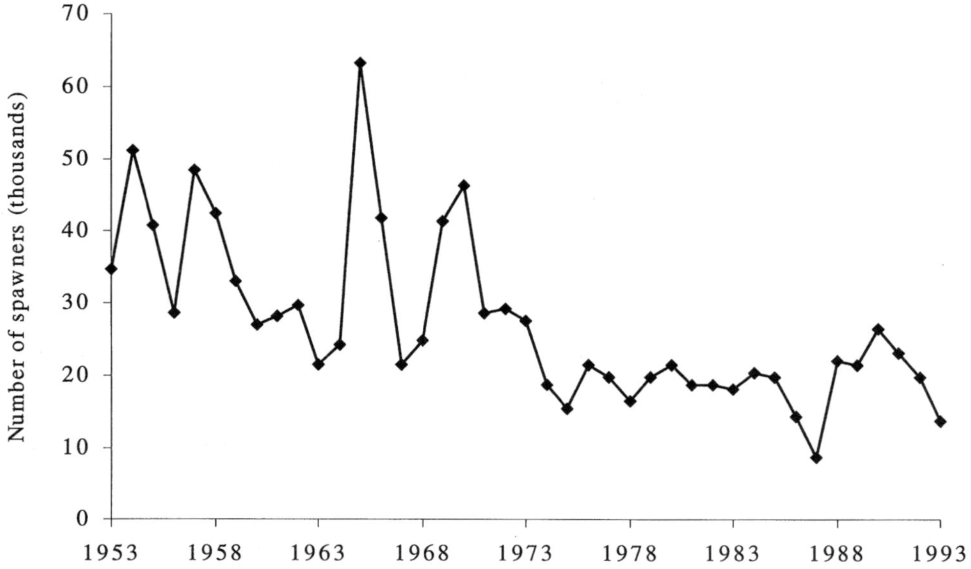

index of abundance does *not* include the substantial catch by commercial troll and sport fisheries in the Strait of Georgia and west coast of Vancouver Island. Catches of these stocks have also declined dramatically in the 1990s. Spawning escapements also decreased in the early 1990s to less than 40 000 fish and these declines have continued up to the present (1998) (Fig. 2.15).

Therefore, while the current status of these chinook and coho populations is difficult to assess because reliable escapement data are limited to only a few wild stocks, it is apparent that these populations have experienced significant declines since the 1970s. During the same period, the proportion of the total escapement of chinook and coho that represents hatchery-reared salmon has grown significantly. The same general pattern of declining escapement is seen for coho salmon in northern BC on the Skeena River where recent escapement indices are one-third the level that they were in the 1960s. This decline has been generally paralleled by a decline in catches.

An important feature of Canadian Pacific salmon is that the average body size at a given age has been decreasing for many years (McKinnel 1995), leading to decreased economic value per fish and reduced reproductive potential. Reasons for this change are discussed in Chapter 3.

Arctic Canada fisheries

The Canadian Arctic sustains three competing fisheries: domestic, sport, and commercial. Each of these operates in freshwater and marine systems (see summary in Crawford 1989). The domestic fishery refers to the aboriginal subsistence fishery (personal and family consumption) that is unlicensed and unregulated. Fishing policies generally recognize that the domestic fishery takes precedence over the commercial or sport fishery. Crawford (1989) considers that possibly the "most significant unresolved aspect" of

Fig. 2.15. Time series of total abundance and escapement of spawners for coho salmon in the Strait of Georgia region. Estimated river spawning escapements and terminal commercial and Native net catches for Fraser River. Data include wild and enhanced river spawners only (i.e., hatchery broodstock and hatchery surplus excluded). Data from DFO (1996*c*).

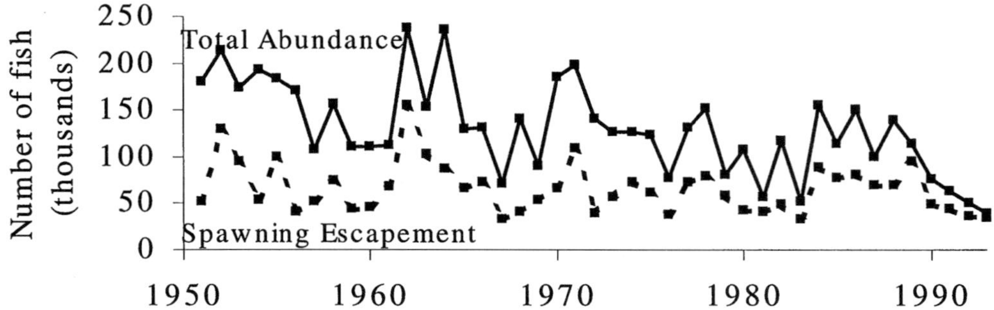

northern Canadian fisheries is the paucity of catch or landings estimates from the domestic fishery. Landings in this fishery may be decreasing due to global change (changing economy and increased interaction with southern regions), although this possibility remains uncertain (few or none data). The arctic sport fishery involves residents and non-residents and though effort has increased in recent years "the potential for growth of the sport fisheries is real and likely to be realized" (Crawford 1989, p. 16). The arctic commercial fishery is mainly an export fishery of anadromous species, primarily Arctic char on the west coast of Hudson Bay and in Cambridge Bay. The freshwater fishery is primarily whitefish in Great Slave Lake and in the Mackenzie Delta region where char is anadromous but fished in freshwater. Commercial catches in the Northwest Territories in 1989 were approximately 50% of their 1950 maximum and in 1989 represented 3% of the total Canadian commercial fishery — a considerable proportion when viewed as a source of income in a relatively low-population region with a large subsistence culture.

The arctic marine fishery is primarily a commercial/exploratory and domestic fishery and is concentrated in the eastern Hudson Strait and Baffin Island regions. Turbot (Greenland cod) landings in the eastern Northwest Territories increased dramatically during the period 1988–1993 (DFO 1995*a*). Most of the ~135 arctic marine species (McAllister et al. 1985) are generally unexploited relative to freshwater species, although that appears to depend on local communities and tradition. The contribution of marine fish to the domestic harvest is generally unknown and many marine species (e.g., fourhorn sculpin, Pacific herring) are represented in the by-catch of the Arctic char fishery. In contrast, northern shrimp support a large commercial fishery, especially in the Davis Strait. Striped pink shrimp is landed as bycatch and a fishery is being developed and managed by the Northern Shrimp Advisory Committee (NSAC).

Marine invertebrate fisheries

There has been a generally steady increase in landings of crustaceans and molluscs in Canada from 1961 to 1993 (Fig. 2.16). The major species are American sea scallop, queen (snow) crab, shrimp (northern and pink), American lobster, and soft shell clam. Lobster landings almost tripled through the 1980s (Fig. 2.17) and reached a maximum in 1990 (46 433 tonnes). This increase is attributed to increased effort and better technology as well

Fig. 2.16. Time series of reported landings (catch) by weight from the Canadian invertebrate fishery in terms of total catch, crustaceans (lobster, shrimp etc.), molluscs (scallops, mussels, etc.), and cephalopods (squid and octopus). The data are derived from FAO (1995*a*).

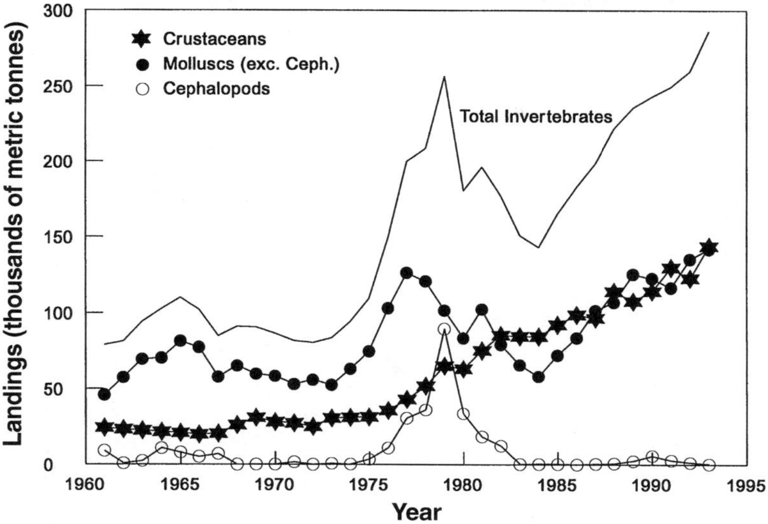

as increased lobster biomass on the fishing grounds that resulted from environmental variation that favored recruitment (Savard 1994). Cephalopod (squid) landings peaked in the late 1970s. Sea scallop and Iceland scallop landings vary among fishing areas and among years and abundance appears to fluctuate significantly with variations in recruitment. Scallops are subject to occasional mass mortalities (e.g., in 1993, 50% died along the lower north shore of Quebec; Giguere and Frechette 1994). Northern shrimp landings in Canada in 1993 were the highest of the preceding five years (FAO) and there has been a steady increase in landings since 1970 for the entire Atlantic northwest region (Stamatopoulos 1993).

Although total inevertebrate landings on Canada's west coast increased fairly steadily throughout the 1980s (Fig. 2.18), this has not been true for the three primary invertebrates which are geoduck, sea urchin, and crab. The geoduck clam fishery in British Columbia has increased dramatically in value since the late 1980s, despite consistent reductions in the total allowable catch; in 1995 the landed value exceeded $42 million. Most of the product is exported live to Asian markets, principally Hong Kong and mainland China. Prices paid to harvesters have jumped from about $2.70/kg in 1990 to about $20/kg in 1995. The annual harvest of sea urchin roe is almost as valuable as the geoduck yield in BC with crab being very close in value. Most of the sea urchin roe is exported to Japan.

Throughout most of its range in the NW Atlantic, catches of the American lobster reached historical maxima in the late 1980s and early 1990s and are thought to reflect increased recruitment in the face of high exploitation rates (Drinkwater et al. 1996). Furthermore, the changes in catch do not appear to be associated with changes in St. Lawrence river runoff as has been previously reported (Sutcliffe 1973). Lobster landings in Atlantic Canada reflect this widespread pattern, which has been paralleled by an increased value of the fishery (Fig. 2.17). Lobster fishing effort data are very limited and thus the landings pattern is a very uncertain reflection of the state of the lobster stocks. Drinkwater et al. (1996) acknowledge this important fact and point out that effort (numbers of traps hauled, advanced technology, and fishing further afield) has increased in recent years over most of the

Fig. 2.17. A time series of lobster landings (as a histogram in tonnes) and value (as a line in millions of dollars) for the the Scotia Fundy region of Atlantic Canada. The data are derived from DFO (1993), DFO (1996*e*), and FRCC (1995).

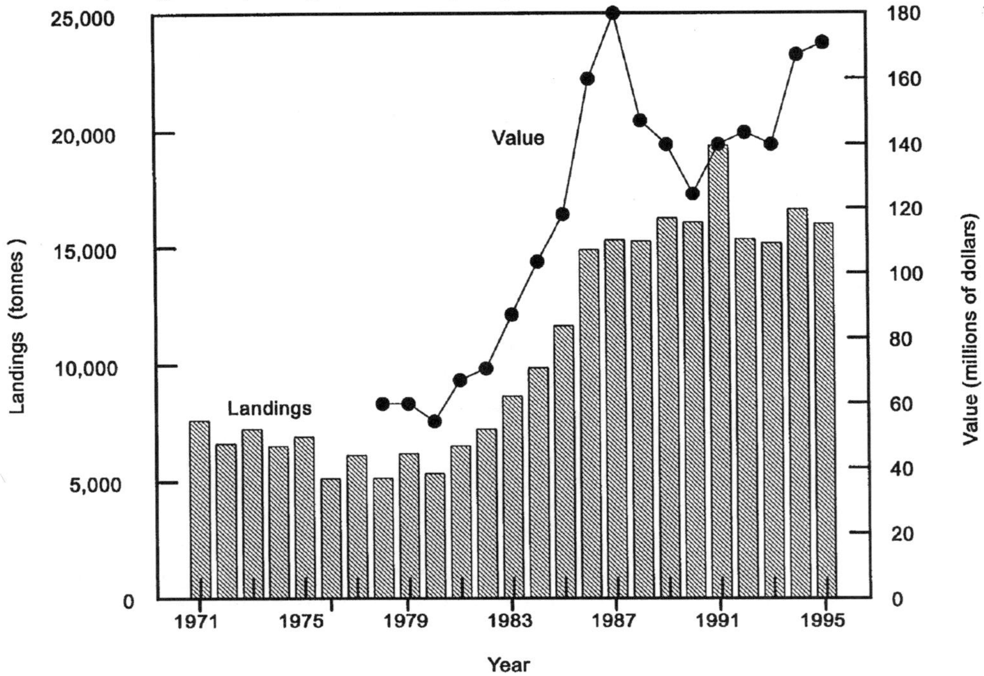

region and that "increased catch rates may be due to the use of more efficient and larger traps and better navigational gear." However, other aspects of the fishery and monitoring data (e.g., research surveys) suggest that the increases in catch do reflect increases in abundance. Nevertheless, the landings data as illustrated in Fig. 2.17 provide no uncertainty estimates and the basic assumptions for drawing conclusions about the ability of catch data to reflect stock size have not been fully met. Thus, maintaining current catch rates at contemporary levels may be highly risk prone.

Fig. 2.18. Landings of invertebrates in tonnes in British Columbia, 1981–1993.

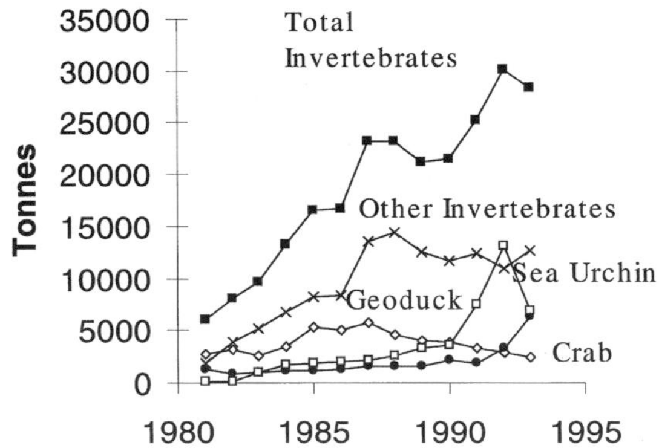

Box 2.5. Aquaculture and Mariculture

The term aquaculture is used to refer to both freshwater and saltwater farming of plants and animals, whereas mariculture refers only to saltwater farming. Canada has moved slowly in its development of aquaculture industries, partly because of the low per capita fish consumption rate in Canada relative to many other countries, harsh climate, and, until recently, relatively high production from the capture fisheries (Boghen 1995). However, aquaculture in Canada is changing dramatically. Between 1984 and 1992 production in Canada increased 300% relative to 27% in the USA. (Overall, Canada increased its share of North American production from 2% to 19% over that period (FAO 1995b). Atlantic Canada has seen a sharp increase in volume and value of aquaculture products from 1986 to 1993 (Boghen 1995). The majority of eastern Canadian aquaculture is mariculture, 87% in 1987 through mariculture and 13% through freshwater aquaculture (rainbow and brook trout, primarily in Quebec). From a taxonomic and regional perspective, aquaculture on the Pacific coast of Canada is dominated by salmon (mostly Atlantic and chinook salmon) and some marine molluscs (mainly oyster). In 1993 this region accounted for 54% of the total Canadian aquaculture production (DFO 1995*b*). Central Canada (including Quebec, Ontario, Prairies) accounted for the remaining 9.6% of the total in 1993 and was comprised primarily of rainbow and brook trout and negligible amounts of shellfish and Arctic char (DFO 1995*b*). In 1993 Atlantic Canada (NB, NS, PEI, NF) aquaculture was dominated by the production of Atlantic salmon, blue mussels, European and American oysters, and negligible amounts of scallops and cod, and accounted for 36.4 % of the total Canadian aquaculture harvest (DFO 1995*b*). In comparison, the global aquaculture harvest increased steadily from 1984 such that in 1992 more than 14% of all fisheries harvest by weight came from aquaculture (FAO 1995*b*). In 1984 this proportion was only 8.3%. The most important aquacultural countries or regions are China, India, and Indonesia. During 1992, Asia was responsible for 84% of the global aquaculture production, while North America accounted for just ~3.2% of the total. It is expected that the global demand for fishery-based products will increase at a faster rate than their availability. Thus, it may be "a serious mistake to propose aquaculture as a panacea for the world's future food shortages" (Boghen 1989, p. 5). One aspect of this dilemma is that fish are now being caught to feed other fish, i.e., instead of directly consuming all the fish that are caught, we are reprocessing them into food for more marketable fish or "value-added" products. Total aquaculture harvest in 1994, a record 25.5×10^6 tonnes of finfish, shellfish, and aquatic plants, was valued at US $39.83 billion. Freshwater culture accounted for 63% of the harvested weight relative to 29% from mariculture and 8% from brackish water. However, the latter represented 22% of the value and resulted primarily from the culture of ~863 000 tonnes of shrimp species.

The world harvest fisheries based on catch statistics

According to the Food and Agricultural Organization (FAO 1995*a*) the world harvest of fish and shellfish in 1994 from the capture and aquaculture fisheries, marine and freshwater (Fig. 2.19) reached a record level of 109.6×10^6 tonnes (see also Garcia and Newton 1994). The quantity used for human consumption amounted to ~68% and the remaining 32% was used to produce byproducts (greatest proportion to date used for such purposes). The recent rise in catch is primarily due to increased catches of anchoveta off Peru and walleye pollock in the North Pacific. Catch does not necessarily reflect the total abundance

Fig. 2.19. Time series of the world total fish catch (millions of metric tonnes) in terms of demersal and pelagic marine species, crustaceans, molluscs, and cephalopods, and freshwater and diadramous species. The data are from FAO (1998).

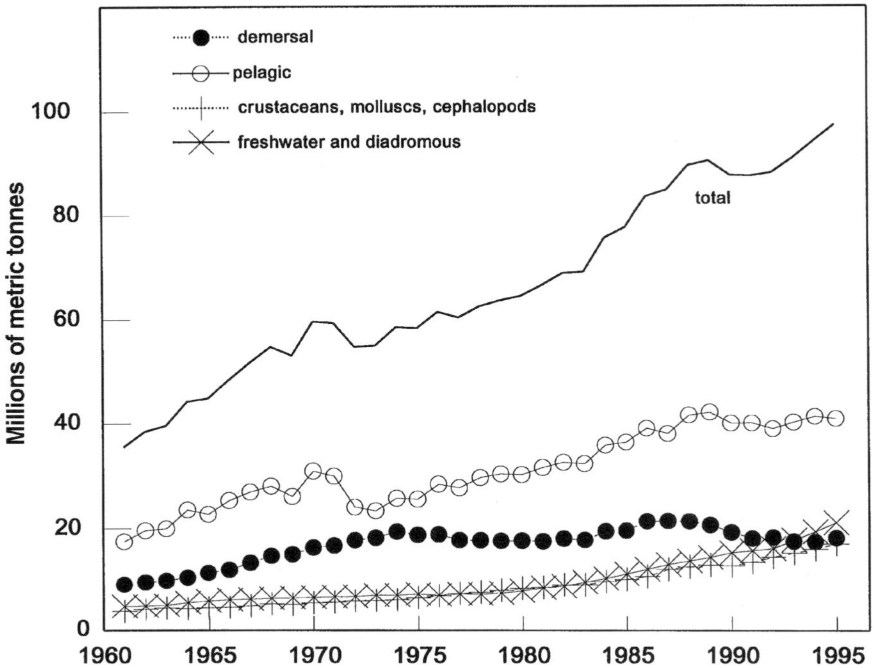

of the species being harvested. However, there is absolutely no question that the world harvesting capacity is enormous, and therefore in general, constant or declining harvests from wild fisheries are most easily, though not necessarily, explained by fishing having met or exceeded the renewable capacity of the wild fish being harvested. *The face of the world fisheries is changing — harvest rates of most wild species have either reached or exceeded their sustainable maxima or are declining, while harvests from aquacultural practices are steadily increasing. Thus, the overall state of the Canadian fishery is not unique in a global context.*

The commercial value of fisheries

The fishing industries in Atlantic and Pacific Canada are export oriented and yet their products represent only a very small percentage of the total world fish market. The Canadian industries therefore have no control over total global harvests and they are subject to periodic economic crises. They also suffer from *chronic overcapacity.* In detail, each of features these can be elaborated as follows.

- They are export oriented. More than 80% of the industry output in value is sold outside of Canada. The industries are "price takers" in that they exercise little or no market power. The prices they receive for their products are determined by supply and demand forces in international markets.
- They have no control over total harvests. The harvests are subject to natural fluctuations in availability and catchability, as well as to shifts in the policies of regulatory agencies in other countries.
- They are subject to periodic economic crises related to natural fluctuations in fish

resources and the compounding nature of chronic excess capacity. The Atlantic Canada groundfish fisheries experienced two major crises (1973–1974 and 1981–1982) within the twenty-year period immediately preceding the current crisis. Each of these crises was severe enough to require substantial financial and social assistance from the federal government. The west coast fishing industry is currently experiencing economic difficulties arising from the state of the salmon fishery and, as on the east coast, such economic crises are not new. In 1982 Canada's Pacific fisheries were "at a crisis point" and "following two depressed years, the economic circumstances of the commercial fishery are exceptionally bleak" (Canada 1982*b*).

- The Atlantic and Pacific coasts suffer from chronic overcapacity (i.e., fishing power) which, despite several commissioned reports, the regulatory agencies have found frustratingly difficult to resolve. In British Columbia, an elaborate limited entry/buyback scheme was introduced in 1969 to address the massive overcapacity problem in the salmon fleet. By 1995–1996, the number of vessels in the fleet had been reduced by 45% (Fig. 2.20). The success is illusory because the number of seiners, the most powerful fishing vessels, actually increased over the same period (Fig. 2.20). Furthermore, the fleet is much more technologically advanced and more efficient than it was 25 years ago. In 1995, recognizing that serious overcapacity remained, it was recommended that the fleet capacity should be reduced by 25–33%, over a 10-year period (Canada 1995*a*). This led to the introduction of the highly controversial "Mifflin Plan." In Atlantic Canada, one of the major themes of the Kirby Report (Canada 1982*a*) was chronic excess capacity in both the harvesting and processing sectors of the groundfish industry. There clearly has been a reduction in capacity on the east coast, with plants closed and ships sold, but there remains concern that overcapacity remains relative to the potential of any future sustainable fisheries (Cashin Report 1993; Schrank 1996).

Fig. 2.20. The number of Pacific salmon fishing vessels by gear sector at various times during the 1966–1996 period. The data are derived from Fraser (1977), Gislason (1996), and PSA (1997).

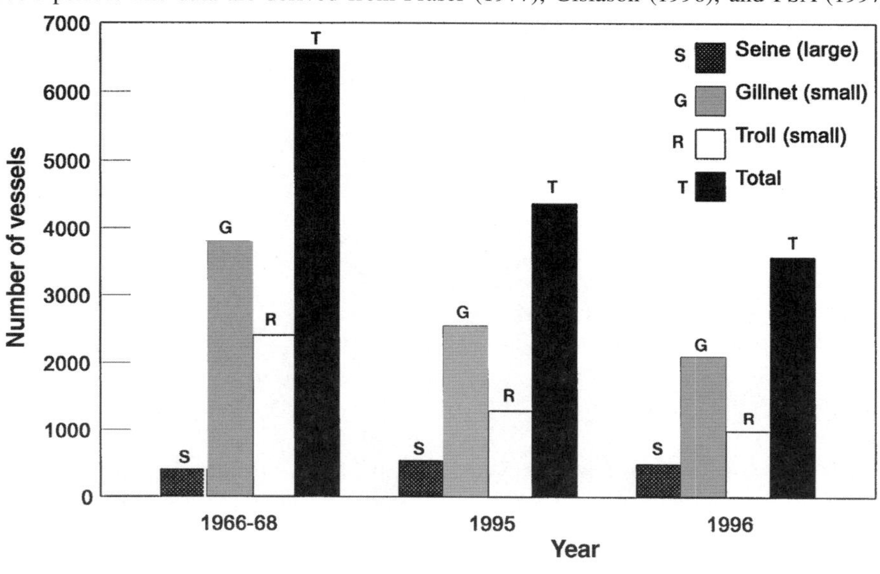

Box 2.6. A Perspective on Fisheries Economics

Capture fisheries in Canada and elsewhere have proven to be extremely difficult to manage effectively in economic terms. *Economists now view fishery resources as a form of "natural" capital — natural resources capable of yielding a stream of economic return to society through time* (Clark 1990; Clark and Munro 1994). Fishery resources and other natural resources, such as capital, differ from "conventional" or human-made capital in that the initial endowment of the asset is received directly from nature. However, just as it is possible to invest or disinvest in conventional capital, it is possible to invest or disinvest in natural capital that exists in the form of fishery resources. *Refraining from harvesting or engaging in activities that conserve or enhance the fishery resource are acts of resource investment. Conversely, harvesting in excess of some desired sustainable yield or destroying habitat serve to deplete the resource and are acts of resource disinvestment.*

The economic motivation for natural resource investment is essentially the same as it is for investment in conventional capital. Such an investment requires a current sacrifice, e.g., reduced harvesting. Having made the sacrifice (the investment), the expected return is an increased resource "capital" or stock and it is hoped that the enhanced stock will yield a greater or more stable stream of economic return in the future. Since the return from the investment is to be enjoyed only in the future, it necessarily means that the resource investment, as with other forms of investment, must be considered with a degree of uncertainty, sometimes extreme. Unfortunately, most individuals and societies give less weight to future returns than they do to present returns and they do so for a variety of reasons. That is to say, they discount future returns and the greater the rate of discount the less will be the incentive to invest in any form of capital. Indeed, if the rate of discount is sufficiently high, there may not be any incentive to maintain the existing stock of capital and disinvestment (i.e., stock depletion) will occur. The entire process reflects the first of several powerful but adverse feedback systems.

With this in mind, consider the basic economics of capture fisheries. The key reason given by economists for the poor economic performance frequently observed in capture fisheries is the existence of ill-defined rights (property rights, access rights, management rights, etc.) to the fishery (Gordon 1954). Although coastal states, such as Canada, may claim such rights to fishery resources within their Exclusive Economic Zones (EEZs), they have in the past found it difficult and costly to vest the rights in the resource to the fishermen on either an individual or collective basis (Gordon 1954; Munro and Scott 1985). The fact that fish are mobile and not readily observable prior to capture explains some of the difficulty. Nonetheless, *because fishery rights are ill-defined, sometimes called the "common pool" problem, the economic incentives for the participants are perverse, i.e., away from what is right or good for the sustainability of the resource. It is this perverse economic incentive system, rather than "greed" per se, that lies at the heart of the economic problems plaguing so many capture fisheries.* There are two major consequences.

The first arises from the fact that without clearly defined rights individual fishermen have an incentive to discount future returns from the fishery resource. Consequently, there is no incentive to invest in the resource and to protect the future returns. If, in such a system, some, but not all, fishermen attempt to invest through reducing their harvest, their actions serve to increase the harvest of their competitors. Thus, *instead of having an incentive to invest in the resource, fishermen have an incentive to actively disinvest. Disinvestment leads to chronic overexploitation of the resource, i.e., excessive depletion of the "natural" capital from society's point of view. Overexploitation of the resource is all but inevitable in "open access" fisheries where the property rights are ill-defined and regulations are either absent or difficult to enforce.* The multitude of regulations employed in Canadian fisheries are designed, in part, to limit overexploitation. Obviously, the regulations do not always succeed.

The second major consequence of the perverse incentive system is excess fleet (and likely processing) capacity. If regulatory authorities attempt to guard against overexploitation by placing year-to-year limits on the allowable harvest and little more, then the limited harvest will become the "common pool." Competition among the fishermen for shares of the limited yet valuable harvest creates incentive for increased investment and thus excess capacity when the resource becomes more limited; i.e., the harvesting power of the fleet will exceed by a significant margin that which is required to take the allowable harvest, even when allowing for harvest fluctuations through time. *It becomes entirely rational for an individual fisherman to invest in additional harvesting capacity (vessels or technology), even though such an investment leads to zero additional total harvest*. This whole process reflects a second powerful but adverse feedback system.

Overcapacity develops for many reasons, one of them associated with a concept called "ratcheting" (Caddy and Gulland 1983). During good economic times there is an incentive to expand harvesting capacity because there is an immediate economic return. As the capacity increases, thereby often leading to increases in catches, the fishing resource develops status and profile and therefore value. When the decline occurs, as it inevitably does, government regulations tend to support the fishery, thereby mitigating the negative effects of the decline. After the return of the fish, the cycle begins again and the harvesting capacity that remains more or less unchanged continues to increase again. This process reflects a third powerful but adverse feedback system.

The emergence of excess capacity results in economic waste and serves to undermine the economic viability of fishery systems. Furthermore, the more that fishing industries suffer from excess capacity, the more vulnerable they become to adverse resource shocks and to adverse shocks arising from the economy (e.g., rising costs and falling prices). Adverse sociopolitical shocks inevitably follow. Excessive capacity also exacerbates the monitoring, control, and regulatory problems of the resource managers. It can be argued that when resource managers are compelled to set harvest limits under conditions of uncertainty, the presence of a large chronically underutilized capacity will lead to pressure being placed on the resource managers to set excessively liberal limits (Dupont 1996). This reflects a fourth adverse feedback system.

Economics of the west coast fisheries

The capture fisheries of British Columbia have been, until very recently, economically dominated by salmon (Fig. 2.21). The crisis of the early 1980s was followed by years of generally increasing value of salmon harvests, with a maximum being reached in 1988. Indeed, the value of salmon landings (Fig. 2.21) in real terms was ~40% greater than it was five years earlier. The volume of salmon harvests began to decline in the mid-1980s, but the decline was, for a time, masked by the rising value. Since 1988, the decline in harvest value has been precipitous. The rapid decline in landed value was not due to declining landings alone because the value per tonne (in real terms) of salmon harvests in 1995 was 40% of what it had been in 1988. The unit value is a function of shifts in prices and shifts in species

Fig. 2.21. Real landed value of Pacific fisheries.

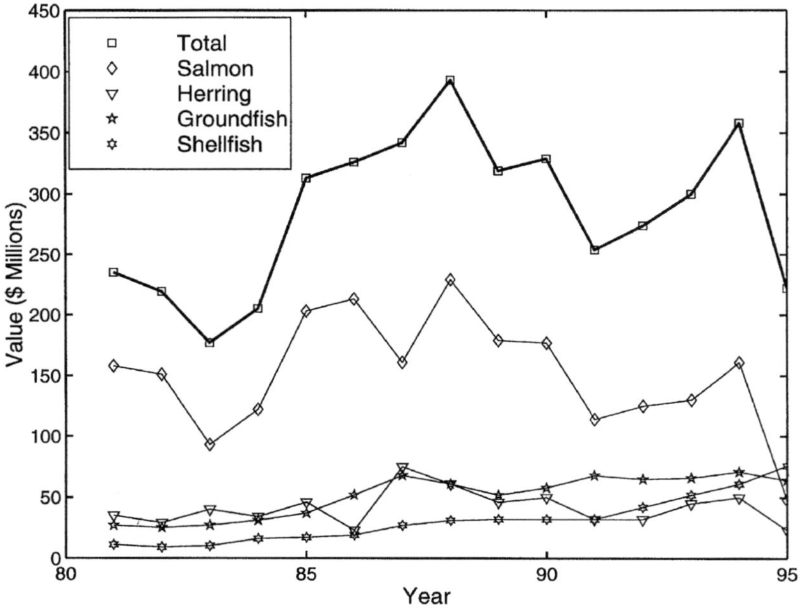

composition (e.g., price of chinook salmon can be six-fold greater than that of pink salmon). In summary, total dollar value of landings in British Columbia fisheries increased steadily from the crisis years of the early 1980s until the value of the salmon harvest peaked in 1988 (Fig. 2.21). While the total value has decreased in most years since then, the value of invertebrate fisheries on the west coast has increased.

Economics of the east coast fisheries

The collapse of the groundfish fishery in Atlantic Canada, with northern cod providing the most spectacular example, has been of overwhelming economic importance. Groundfish harvests in 1995 were only 11% of what they had been in 1982 (Fig. 2.22). Although the groundfish fisheries have been severely affected economically by the resource collapse (Gordon and Munro 1996), there has not been a precipitous decline in the total dollar value of all landings in Atlantic Canada (Fig. 2.23). This is because of the growing value of shellfish and crab (Fig. 2.23). Traditionally, one used to think of the fish harvests in Atlantic Canada as being dominated by groundfish in value, as well as in volume. While groundfish did indeed account for almost one-half of the landed value in 1981 (Gordon and Munro 1996), their value was in fact only marginally ahead of that for shellfish and crab. A comparable examination in 1995 shows the value of shellfish clearly dominating, a situation that is very similar to that seen for the west coast of Canada.

While there have been fluctuations in the landed value of harvests in Atlantic Canada from 1981 to 1995 in real terms, overall there appears to have been a gradual upward trend (Fig. 2.23). Indeed, in 1995 the landed value (in real terms) was roughly two-thirds greater than it had been in the early 1980s, in spite of the groundfish collapse. Of the shellfish harvests, crab show the most striking increase in recent years. Between 1990 and 1995, the landed value of crab (adjusted for inflation) increased by 750% and again, the increase was

Fig. 2.22. Atlantic fisheries catch trends.

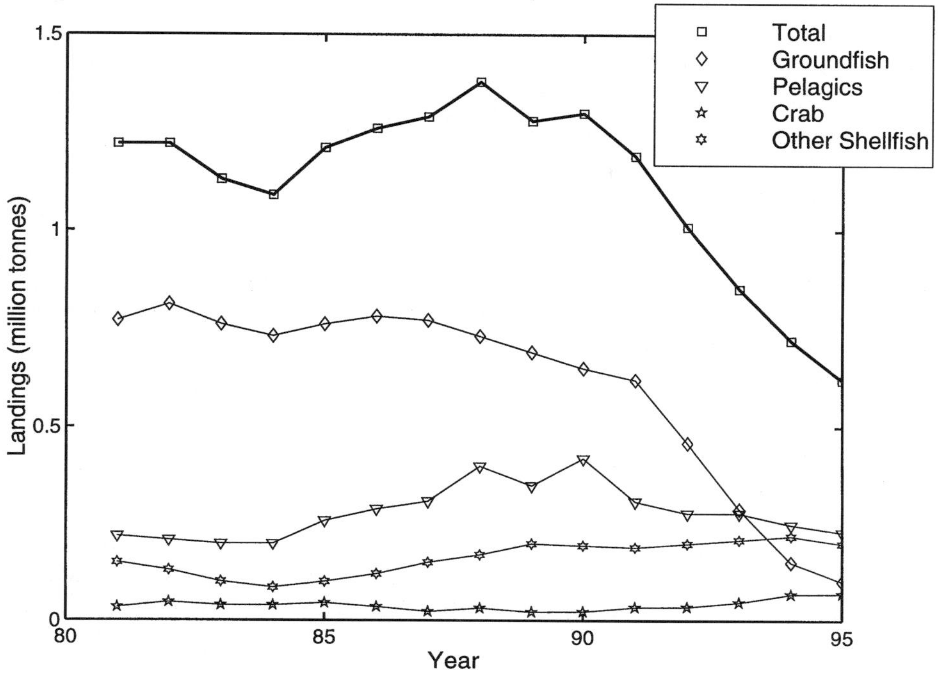

Fig. 2.23. Real landed value of Atlantic fisheries.

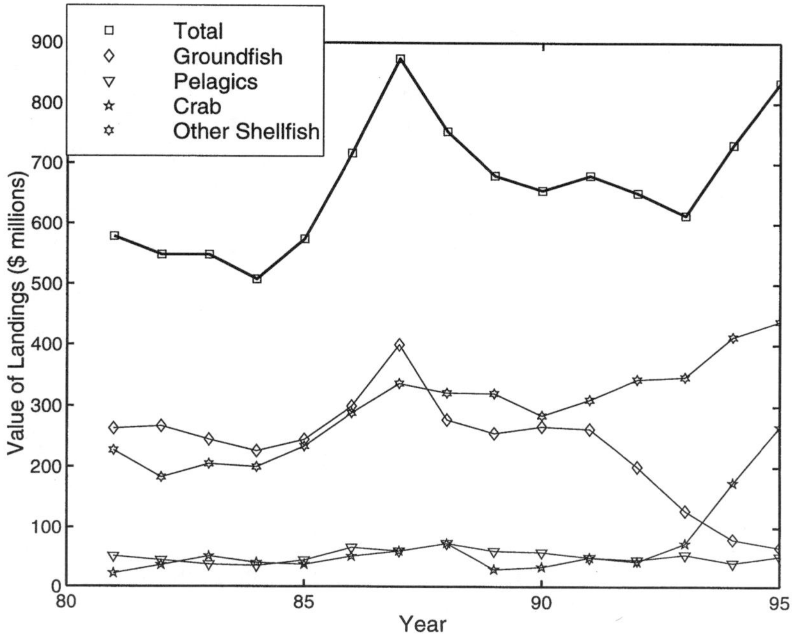

due to a combination of increased landings and rising unit value (in real terms). The question remains, of course, whether the shellfish harvests, crab in particular, are sustainable at these levels.

One aspect of the Atlantic fisheries, which may well have served to exacerbate the overcapacity problem, is the extensive federal and provincial subsidy programs, e.g., funding for corporate processors, boat-building loans, and unemployment insurance (a program which in late 1996 underwent substantive changes). While subsidies are used on the Pacific coast, they are not of the same magnitude as those in Atlantic Canada. It is difficult to separate subsidies from government expenditures for resource management purposes (which some have argued constitute an indirect subsidy to the industry). However, over the decade of the early 1980s to the early 1990s, total federal and provincial government financial outlay on Atlantic fisheries was conservatively estimated to be in the order of $8 billion, half of which was accounted for by unemployment insurance. The outlay of $8 billion was equal to the gross value of landed harvests in Atlantic Canada over this period (Schrank 1996), which might approximate half of the total value after processing.

Economics of global fisheries

The chronic economic difficulties experienced by Canada's Atlantic and Pacific fishing industries are not unique to Canada. Capture fishery resources throughout the world have proven to be extremely difficult to manage effectively in economic terms. This is a reflection of the "tragedy of the commons" or the common pool problem that afflicts these fisheries (see Box 2.6).

The United Nations Third Conference on the Law of the Sea 1973–1982 attempted to mitigate the "tragedy of the commons" through the establishment of 200 mile Exclusive Economic Zones for coastal states like Canada and Extended Fisheries Jurisdiction (EFJ). EFJ transformed large amounts of fishery resources from international common property to coastal state property. The results of EFJ worldwide have, to date, been disappointing. First, there remains ample evidence of continued overexploitation of the resources. Using FAO estimates, Garcia and Newton (1994) report that almost 70% of the world fishery resources are either fully exploited, overexploited, depleted (i.e., commercially extinct), or are recovering from past depletion. Garcia and Newton (1994) note that, given the state of present resource management in the world, the fully exploited resources are primed to become overfished or depleted resources. If anything, the global statistics underestimate the degree of overexploitation in world capture fisheries. Garcia and Newton (1994) further estimate that 30% of the existing worldwide fleet capacity constitutes excess capacity and that the excess capacity problem has been exacerbated by the widespread subsidization by governments of the fishing industry. The overall economic performance of world fishing industries based on capture fisheries has been dismal. The FAO estimated that the 1992 total landed value of world harvests of marine fish to be US $70 billion. FAO further estimated that fleet operating costs alone exceeded the landed value by US $22 billion. If one includes capital costs, the losses amount to US $54 billion (FAO 1995*a*). Losses of this magnitude imply substantial subsidization of fishery activities on a worldwide basis.

The socio-political diversity of Canadian fisheries

The participants in any fishery system have agendas that reflect a diversity of knowledge, opinion, interests, and objectives. The socio-political characteristics of these participants and their geographic distribution have implications for how management problems are addressed and how policy options are conceived and implemented. Generally, the participants vary according to the following non-exclusive criteria: regional representation, ownership or non-ownership of resources and processing, the degree of capitalization, dependence on capital supply, and political power and influence, vis-à-vis the regulatory agencies. The most significant direct participants include: aboriginal fishermen, a variety of commercial fishermen, processing companies, shoreworkers, recreational fishermen, the federal and provincial governments, non-governmental organizations (NGOs), and fish-dependent communities.

Commercial fishermen

In 1991 there were more than 37 000 full-time commercial fishermen in Canada (Cashin 1993). If the number of part-time fishermen who earn a significant portion of their income from fishing were included the total would approximately double. Small-scale inshore fishermen in the Atlantic provinces (NF, NS, NB, PEI, PQ) tend to be self-employed owner–operators who earned between $17 000 and $26 000 in 1990 (Cashin 1993). This amount is greater than that for fishermen wage earners and 17–25% less than the average Atlantic Canada wage earners in all employment sectors.

On both coasts the majority of commercial fishing units has been, and still may be, small-scale and owner-operated. However, the relative number of the larger and more capitalized fishing vessels has been increasing, as has their allocation. As the capacity, capital cost, and technological sophistication of the larger vessels increased over the last 30 years, they became either more indebted to major fish processing companies or partially or wholly owned by them. In BC, for example, the larger seine vessels (usually over 75 ft; 1 ft = 30.48 cm) have more than doubled as a proportion of the salmon fleet from ~6% in the 1960s to ~14% in 1996, (mostly at the expense of the small boats in the troll fishery) while small gillnetters have remained a relatively constant proportion at ~58% (Fraser 1977; Gislason 1996; PSA 1997). This trend may continue in response to ongoing fleet reduction policies.

In Maritime Canada the total number of fishing vessels has remained surprisingly constant at around 11 000 over the 1971–1991 period (Cashin 1993). Over this period the small vessels (generally less than 50 ft) consistently comprised 89–95% of the total while midsized vessels (45–75 ft) comprised 3–4%, and large vessels (>100 ft) fell from 1.4% in 1971 to 1% in 1991 (Cashin 1993).

In contrast to the Maritimes, the number of vessels in Newfoundland almost doubled from 9650 in 1971 to 15 335 in 1991 although a maximum of 17 135 vessels were recorded in 1981 (Cashin 1993). However, consistent with the other Atlantic provinces, the proportional distribution among vessel size-classes remained remarkably constant at 91–93% for small vessels (<35 ft), 7–9% for midsized vessels (35–99 ft), and 0.5–0.7% for large vessels (>100 ft). The growth in inshore (small) vessel licenses was largely attributable to shell fisheries (e.g., lobster and crab) when groundfish licenses were lost to the midshore (midsize) sector (Davis 1991).

In Newfoundland, vessels classified as “offshore” landed about 50% of the groundfish and some 40% of the shellfish in 1991. However, part of the fleet classified as “inshore” is in reality also represented by the more capitalized “matcher” fleet which exploits both inshore and offshore waters. This may, in part, explain why the highly capitalized large-vessel (>100 ft) offshore fleets on the Atlantic coast have decreased in absolute numbers since the 1980s because they have been effectively replaced by the midshore, midsized “matcher” fleet. The growth in licenses and investment in midsized but offshore “matcher” vessels was accompanied by increased allocations to these sectors (Haedrich 1995; Pinkerton 1987; Davis 1991). However, it appears that companies with offshore quota successfully prolonged the Temporary Vessel Replacement Policy (TVRP) by contracting the matcher fleet to catch the offshore quota that could not be caught by the shrinking large-vessel offshore fleet. Thus, inshore and offshore quota categories and small-, mid-, and large-vessel categories no longer seem capable of reflecting the complex dynamics of the various fleet sectors to secure quota.

Processing companies

In 1991 there were over 1000 fish processing plants in Atlantic Canada and 212 plants in BC and there have been reductions in number since then. Nevertheless, during the eight-year period between 1983 and 1991, the number of registered fishing plants in Atlantic Canada almost doubled from ~670 to ~1063 (Cashin 1993). Both the Atlantic and Pacific fisheries are dominated by major multinational companies which are the principle buyers, processors, and marketers of most, if not all, species. There are two major companies on the east coast and two on the west coast. Although there are also many small-scale processing plants, they tend to operate on the margin and occupy market niches left open by the major corporations. It is interesting to note that over the ten year period, from 1981 to 1992, when fish catches on the east coast declined substantially, the number of fish plants there actually increased (Fig. 2.24).

To ensure ready access to markets, fishermen normally establish a functional relationship with processors. The processor is motivated to regularize this relationship in order to guarantee supply. A processor may even avoid price competition if a fisherman is indebted

Fig. 2.24. The number of DFO registered fish processing plants in Atlantic Canada during the period 1981–1991. The data are derived from Cashin (1993).

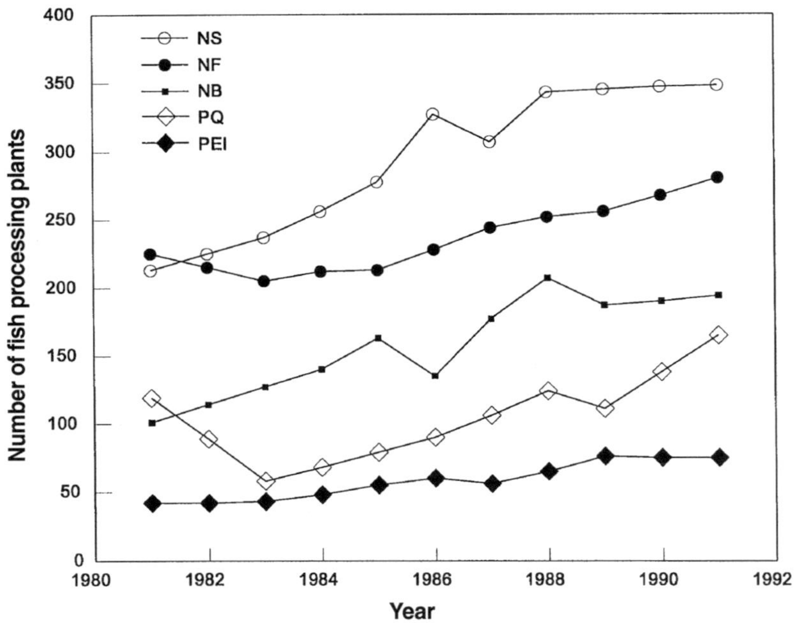

to the processor because the debt can include an obligation to deliver all the catch to the debtor, regardless of better prices offered elsewhere. On the west coast, this form of vertical integration, or "non-price competition," is supplemented by a controlled market for raw salmon and herring roe.

The trend toward increased investment in more efficient vessels in some fisheries is associated with an apparent increase in corporate ownership and (or) corporate mortgage holding in the fleet sector. It is difficult to assess this change because of the predominance of private "handshake" transactions. On the west coast as much as 58% of the seiner fleet is estimated as being owned by, or indebted to, processing companies. On the east coast, there is increased corporate involvement in the offshore and matcher sectors (Davis 1991). Thus, the Canadian fisheries are becoming characterized by an increasing dependence of fishermen on processors. Further, it appears that a relatively small number of fishermen own a large number of multiple licenses in multiple fisheries and these have been gained by the trafficking of limited entry permits and Individual Transferable Quotas (ITQs).

Major fish processing companies are organized into associations on each coast and nationally in the Fisheries Council of Canada, which has a major influence on policymakers. Fishermen have unions and associations on both coasts perhaps with less influence because no single body represents the majority of fishermen. The variety of associations and unions reflects the diversity of the participants. The Canadian Council of Professional Fish Harvesters has become more important in recent years as a national body.

Box 2.7. Social Structures and Fisheries Systems

Fisheries management has historically placed emphasis on capacity and capitalization as drivers in the development of a fishery. Thus, overcapitalization or the development of excess capacity can be driving forces in the creation of fisheries problems. The tendency to overcapitalize is not randomly distributed throughout a fishing fleet but in some cases can be concentrated in the larger vessel and corporate sectors. Small-boat fleets, particularly on the east coast, clearly exhibit overcapacity in any particular fishery, but this overcapacity can be viewed as an adaptive strategy that permits the operators to engage in multi-species fisheries. Can this overcapacity be constrained by a combination of a long-term orientation toward, and understanding of, adjacent marine territories by informal community regulation and by effective regulation and oversight? The answer is neither clear nor easy.

Pinkerton (1989) shows that small-scale owner–operators (relative to corporate large-scale operators) tend to view fishing as: (1) a multi-generational occupation in which knowledge, skills, and ownership are transferred within families and communities; (2) a local or place-oriented activity that is dependent on the health of local resources, habitats, and ecosystem with which there is historical familiarity; and (3) an occupation which combines aspects of commercial and artisanal activity. In addition, small fishing vessels are used for multiple fisheries and for non-fishing activities. Community-oriented values and events frequently take precedence over purely economic (i.e., fishing) activities. On the west coast, these tendencies are particularly salient in the ~30% of the commercial fleet that is aboriginally-owned and operated.

In many preindustrial and small-scale modern fisheries, the local communities managed stocks sustainably (Schlager and Ostrom 1993). Self-management at the local level is as dependent on rules and responsibilities as is governmental management, but they differ in the process by which the responsibilities are recognized, by which rules are made, the consequent legitimacy of the rules, and their enforceability. The willingness and ability of individuals in fishing communities to invest in the establishment, monitoring, and enforcement of local fishing rules creates a form of "social capital" (Ostrom 1990). Social capital may be thought of as the achievements made by a social group whose value can be used to resolve ongoing management problems and conflicts. It can include agreement on rules for sustainable management, conflict resolution mechanisms, agreement on membership, and agreement on what constitutes valid information on the status of the resource. More importantly, social capital enables a social group to create new values — a group within which sufficient trust (new capital) can be used to accomplish much more.

Economic activity within a social group is not independent of social interactions within the group. Societies function according to norms, values, and rules that bind them and at the same time shape their economic choices. Reliable predictions about human behavior are unlikely to be formed simply from an examination of the so-called rational choices of individuals functioning as if they had no history, no culture, and no social reference group whose support and approval are valued. *Management systems that view social capital and social institutions as valuable resources are more likely to succeed than those that do not.* When small-scale fishing societies or postindustrial committees of local fishermen develop the ability to reach agreement at a community level on the who, what, when, and how of fishing, they usually inhibit overcapitalization and avoid gear and allocation conflicts (Pinkerton 1989; Schlager and Ostrom 1993).

Shoreworkers

In 1993 on the west coast, there were more than 6000 shoreworkers employed among 198 fish processing plants in 1993 (provincial estimates). These numbers represent a near three-fold increase over the average of 2400 (Guppy 1987) from the previous decade. This situation contrasts sharply with the east coast where shorework and fishing employment were reduced by about 40 000 between 1991 and 1995, mostly in response to declines and closures in the groundfish fishery. The North American Free Trade Agreement softened Canada's requirement that Canadian landings be processed in Canadian plants. In addition, major Canadian processors have moved some operations to the US (e.g., BC Packers and the Canadian Fishing Company's major cannery in Ketchikan, Alaska). Nonetheless, shoreworkers in fish processing plants constitute the major labor force in many fish-dependent communities, particularly on the east coast, where fish plants are more widely distributed. On the west coast, the major processors and employers have concentrated their operations in Vancouver and Prince Rupert. Only in Prince Rupert do shoreworkers constitute a major portion of the local work force.

Sport fishermen

The growth of sport fishing in the last 10 years is an important trend on both coasts, although it is more prominent on the west coast. Sport fishing lands less than 10% of BC's salmon but it is beginning to challenge the commercial fishery for greater access to several species, particularly chinook and coho salmon.

The sport fishery has two constituents who perceive their interests and allegiances quite differently. The environmentally and locally oriented group involves local associations who wish to improve regulations for conservation and habitat restoration. They often work co-operatively with other sectors on the local level to help develop selective commercial and sport fishing strategies as well as to restore habitat or stocks. The Steelhead Society of BC is seen by some as the more militant group. The other constituent is a more "industrial" sport fishery represented by charter boat operators, large-scale commercially guided luxury lodge operators. This constituent tends to favor large-scale hatcheries for stock preservation and has at times resisted government monitoring. The sport fishery in BC has encountered fierce opposition to expansion from aboriginal and non-aboriginal coastal communities alike, as well as from the commercial harvest fishery. For example, the Haida Nation and local commercial fishermen of the Queen Charlotte Islands have blocked access to fishing lodges and have attempted to establish local hiring rules and local "finnage" fees in exchange for sport access to traditional fishing areas, commercial species, and limited anchorages.

In contrast, Quebec's growing freshwater sportfish local management programs, comprising Zones d'Exploitation Controllees (ZECs) — operated and funded by sport clubs and overseen by government — have received broad support and appear to contribute to sustainable management. Areas managed by ZECs generally show healthier stocks than elsewhere. ZECs have successfully integrated local rural fishermen with more distant urban members from Montreal and Quebec City in joint efforts to improve stock management through involvement in harvest plans, monitoring, and stocks assessments, etc.

Aboriginal fishermen

All three coasts of Canada have experienced increased claims for access to fish resources and for management by aboriginal groups who are not participants in the traditional commercial fisheries. In the Arctic, innovative co-management agreements have been established between the Inuit and the federal government. There most fishery resources are now managed jointly, through such mechanisms as the Nunavut Wildlife Management Board. The Micmacs of the Gaspe Peninsula and the Montagnais of the Quebec North Shore have also reached access and management-sharing agreements with government. On the west coast, several aboriginal pilot programs for sales and management partnerships are now 5–10 years old. The Canadian Supreme Court decision on Sparrow (Regina v. Sparrow, Supreme Court of Canada, May 31, 1990) placed a responsibility on the federal government to consult closely with the First Nations involved in the fisheries. This duty includes ensuring that First Nation rights of access, as protected under section 35 of the Constitution Act, and as interpreted under Sparrow, are not circumvented by other users before the salmon reach their river of origin. A strict interpretation of this judgment would mean that salmon must now be managed from the river mouth outward rather than inward, a next to impossible feat given the migration and traditional harvesting patterns. Although the First Nations have a strict interpretation of the law on their side, they have often found that their goals are more effectively achieved through collaboration with their non-aboriginal neighbours and the traditional commercial fishery, which on the west coast includes ~20% aboriginal participants. First Nations political power and influence are increasing but much negotiation remains as treaty and land claim processes slowly move forward (PSA 1997).

The Department of Fisheries and Oceans

The primary policy-maker and regulatory agency in the marine fisheries in Canada is the federal Department of Fisheries and Oceans (DFO) whose mandate is provided through the Fisheries Act and the soon-to-be Oceans Act. Under the Fisheries Act the department has the responsibility of regulating all marine fisheries, administering licenses, conducting research, collecting and analyzing data, protecting habitat (a function shared with the provinces), negotiating international agreements, and managing many aspects of the oceans' health. There were over 6000 DFO staff in the 1980s but this has been reduced to about 4000 through budget reductions. These fiscal imperatives have also led the department to explore means of recovering costs from fishermen.

The provinces

While fishermen are licensed by the federal government, the provinces license fish plants and are most affected by the employment levels in them. The diverging interests of federal and provincial regulators can lead to a situation such as the one depicted in Fig. 2.24 where the number of plants on the east coast was increasing while the landings were decreasing. The collapse of the Atlantic cod stocks displaced roughly 40 000 fishermen and shoreworkers from the fishery by 1995. This situation clearly points to the need for federal–provincial regulatory coordination.

In the Atlantic provinces and Quebec, provincial departments of fisheries play a minor regulatory role relative to that of DFO. This may be changing, however, with the federal downsizing of DFO and with the provinces growing increasingly aware that they are directly influenced by federal management and fleet downsizing policies. The provinces therefore attempt to pressure DFO into viewing fish management policies in the broader context of other federal and provincial policies on employment, health and welfare, Indian Affairs, regional economic development, and even national jurisdiction. This has become apparent on the west coast where the provincial government actively opposed DFO's fleet reduction strategy and procured a Memorandum of Understanding with DFO to work toward sharing responsibilities in fish management — including community-based management options. British Columbia and other provinces have recently created separate fisheries departments or ministries and are beginning to play a greater role in fisheries management.

Non-Government Organizations (NGOs)

In recent years a number of NGOs have entered the policy arena, sometimes signaling global fisheries concerns. Environmental groups such as Greenpeace, the Sierra Club, the Canadian Environmental Defense Fund and the Canadian Oceans Caucus, and networks of fishermen such as the North Atlantic Small Harvesters Association and the International Collective in Support of Fishworkers are examples. Sometimes NGOs demonstrate the increasing role of local and regional networks such as the Coastal Communities Network on the west and east coasts and Women's Fishnet on the east coast. Such groups are pressing for marine protected areas, selective fishing practices, protection of biodiversity, eco-labelling, and equitable distribution of access and benefits. Recently, Unilever and the World Wildlife Fund launched the Marine Stewardship Council to develop global standards for sustainable management. NGOs frequently raise funds, supply negotiation tools and media contacts, and work with communities and fishery organizations on projects that are intended to improve management and ecosystem health. Increasingly they are included in policy forums and local boards.

Fishing-dependent communities

The geographic location of communities is viewed as significant by human ecologists because the communities are close to, or part of, important habitats, nursery areas, spawning grounds, or traditional fishing areas. Hence, there is potential for communities to monitor ecosystem linkages, stock productivity, and fishing activities. In addition, the opportunity for communities to decide on habitat development for non-fish usage (marinas, hydroelectric, log-booming, irrigation, etc.) means that they can play a key role. It matters whether the community has an economic interest in the fisheries, because it will likely influence decisions. In other words, it is not only the communities that are often dependent on the resource, but the resources also depend on the human communities.

As a percentage of provincial population, non-urban fishing communities may not seem very significant, outside of their predominance in Newfoundland, because for each of BC, PQ, NB, PEI, and NS, they comprise less than 25%, some as low as 5%, of the total population. However, the communities are widely dispersed along the coastline and thus offer opportunities for involvement in year-round monitoring and other management activities already explored by aboriginal groups. Most of these communities have a high level of fishing dependence because there are few alternative occupations and those they do have in other resource sectors are usually shrinking. It has been argued that their survival as communities depends on the sustainable management of locally accessible fisheries that are critical to the local or regional economy.

BC has a history of regional inequalities in fishing access, along classic metropolis vs. hinterland geographic lines. Processing plants have become more centralized into two urban centres and it might be considered surprising that remote coastal communities have retained licenses as well as they have. The decline of salmon licenses in representative rural areas of the BC coast is roughly comparable to the decline of licenses in the two metropolitan areas. This suggests that fishing in rural areas has held its own relative to urban areas. However, if it is noted that 68% of the large-vessel fleet is located in urban areas, it is possible that rural areas will be far more affected by fleet reduction policies than will urban areas. Implementation of "license stacking" onto fewer boats has been predicted to reduce the large-boat fleet by about 50% and the small-boat fleet by as much as 80% (PSA 1997). Recent allocation debates among small- and large-gear sectors reveal that changes in the size of each sector can become the basis for arguing greater allocation to the sector least reduced as opposed to ensuring that each remaining fishing unit within the small gear sectors becomes more viable. Regional inequalities are thus induced by fleet reduction policies.

Income support and subsidies in Atlantic fisheries

Fishing in most parts of Canada has two very important attributes: it takes place in relatively isolated rural areas and it is an intrinsically seasonal activity. These points combined imply that most fishermen and processors will not work at fishing year-round and will be lacking alternative employment possibilities during the off-season. A special form of Unemployment Insurance (so called "Fishermen's UI," now renamed as Employment Insurance) was established to accommodate the nature of this industry. Unlike regular U.I., which is intended to provide support in the case of an unexpected period of unemployment in a worker's career, Fishermen's U.I. is meant to provide a means to cope with the regular annual ups and downs of the fishery and is used by independent fishermen while regular U.I. is used by those paid wages, in processing plants or on offshore trawlers.

The U.I. benefits received by fishermen and fish processing workers are significant. According to the Task Force on Incomes and Adjustment in the Atlantic Fishery (Cashin 1993; Table 22-1), such benefits provided 30% of the average total family income of "fishing families" in Atlantic Canada for 1990. This varied greatly between provinces, from just over 20% in Nova Scotia to 43% in Newfoundland. For "processing families," there was an average of 27.5%, with a low of 15% in Nova Scotia and a high of 41% in Quebec. (As a comparison, on average U.I. provided 6% of total income for non-fishing families in Atlantic Canada.)

Fishermen's U.I. has a number of effects. By supplementing income from fishing, it enables fishermen (and processing workers) to stay when otherwise they would not. This is a long-term impact — on the negative side, capacity is maintained in an overcapitalized industry, while on the positive side, social stability is enhanced by permitting people to stay in coastal communities. At the same time, by reducing the reliance on fishing to obtain a certain target income (perhaps that needed to provide basic living necessities), U.I. has the beneficial effect of reducing the immediate pressure on the resource. Indeed, given the government's record of setting excessive quotas and an inability to control overfishing, perhaps in the absence of U.I., fishermen would have harvested even more intensively.

Fishermen's U.I. has been the subject of much discussion. In the end, perhaps the definitive statement on the subject came from the above-mentioned Task Force on Incomes and Adjustment in the Atlantic Fishery (Cashin 1993, p.76), which concluded that while changes were needed to update the system, "We believe that the fundamental concept of Fishermen's UI to provide income support during seasonal interruptions of work is sound." We return to this topic in Chapter 3 and again later in the report.

Also much talked about in recent years is The Atlantic Groundfish Strategy ("TAGS") program of income support and retraining for those affected by the collapse of the Atlantic groundfish stocks. This multibillion dollar program has been criticized by some for (*a*) an excessive emphasis on income support, (*b*) maintaining hope for people who (it is said) should be leaving the fishery, and (*c*) for retraining individuals for jobs that do not exist locally, while failing to deal with the fundamental issue of regional job creation. Whatever one feels about this program, it certainly provides a vivid indication of the costs involved in failing to allow for change, uncertainties, and risk in fishery systems, a central theme of this report.

Finally, it is important to note that U.I. benefits and TAGS payments are made to support individual Canadians. On the other hand, considerable government funds have been devoted to supporting boat building subsidies (particularly since the 200-mile limit was declared in 1977) and the restructuring of offshore fishing companies in the early 1980s (which cost governments somewhere between $100–200 million, according to Parsons (1993)). Both of these subsidies helped to create the excess physical capacity that is apparent in many fisheries today.

Box 2.8. A Perspective on Fisheries Institutions

Institutions are organizational arrangements by which people manage themselves. The "market" is an institution, as are fishermen's associations, co-operatives, and municipal governments. This section focuses on how fisheries are affected by the choice and the development of institutions, whether within the fishery or external to it, and whether at the local, provincial, or federal level. Perhaps the most critical aspects of an institutional perspective are as follows. (1) Multiple objectives are common in fisheries systems. The objectives need to be reflected in the choice of institution and the type of management. The objectives must be understood at all levels (fishermen through to the regulatory bodies). Attempts to manage fisheries systems in the absence of clear and explicit objectives are bound to be less than successful. (2) Fishery systems are multidimensional; there is a corresponding need for a comprehensive and "integrated" view that incorporates the ecological, economic, and social components of the fishery as well as the interactions that exist within the broader human and natural environment. To be successful, fishery management must help guide this complex system (including its natural and human dynamics and feedbacks) toward achieving the multiple objectives. If any of the institutional components of the fishery and its environment are ignored in the development of policy, important interconnections among them may be missed, resulting in failed management.

Getting the institution "right" is crucial for successful management. If the institution is well-structured, has wide-spread support, has a fair and just code of conduct, then there is a greater potential for success. Recent institutional arrangements for the conservation of Atlantic groundfish stocks have failed. Some new institutions have been formed as a result (e.g., the Fisheries Resource Conservation Council formed in 1993 to advise the government on conservation). A focus on institutions is common to the social sciences, institutional economics, socio-economic and ecological economics. However, it is relatively uncommon in traditional fisheries economics.

Incentive structures (e.g., fishing rights) and social interactions among fishery participants vary among institutions. There is a role for institutions to focus both on individual incentives and on group dynamics — how people interact in a fishery. On the other hand, the creation of appropriate individual incentives (whether market based or socially based) can be a major challenge for other institutions, such as those charged with developing and enforcing regulations. Thus, a key to an institutional perspective is understanding how institutions develop incentives and how participants react to them.

The institutional structures of Canadian fisheries highlight the limits to management; there remains a belief that more can be controlled than is possible given the complexities and uncertainties found in fishery systems. Only when functional and effective institutional arrangements (along with manageable, enforceable regulations) have been identified can the corresponding financial, administrative, and organizational capabilities be maintained.

This institutional perspective implies an "integrated approach" to addressing policy concerns. It also implies that the "right" institutions may be ones that are not restricted to the fishery alone. Most coastal fisheries face a trio of problems; overexploited resources, overcapitalized fleets and a lack of non-fishing alternatives — fishery policy frequently ignored the latter problem. This is particularly obvious in efforts to address overcapacity by removing "excess" fishermen without solving the problem of where the "excess" may find employment. Such measures tend to merely redirect pressure onto other fishery resources, the same resource through illegal channels, or onto the greater society that is neither directly involved in nor dependent on the fishery.

Marine fisheries management in Canada

In the face of long and short term problems, difficult decisions must continually be made in managing the fisheries. With growing recognition that open-access marine resources must be stringently rationed and regulated to prevent overexploitation, that major stocks are declining, and that there is increasing evidence for the possibility of species extinction. The achievement of sustainability, however defined, has become "the" topic of debate for fisheries management. Evaluated against the sustainability objective, existing management schemes appear to have serious limitations.

Management of harvesting activities in Canada has taken place in a complicated, often overly centralized institutional setting characterized by incentive structures encouraging overinvestment in harvesting capacity, overexploitation of current stocks, and underinvestment in the health and size of future stocks. Practical questions of information flow, organizational structure, enforcement possibilities, and incentives for compliance must be addressed. We must achieve realistic and justified determinations of optimal harvest levels that include estimates of uncertainty.

In Canada, management decisions occur within a federal system in which the responsibility for ocean resources rests with the federal government. In particular, responsibility for the conservation and health of fish stocks rests with the Minister of Fisheries and Oceans, acting normally on the advice from officials of that Department and from arm's-length advisory bodies as requested. The provinces regulate shore-based operations such as fish processing plants and settlement patterns and coastal land use.

The Canadian marine fishery developed differently in dif ferent regions of Canada. The associated management systems, which also developed differently among regions (Gough 1993), are detailed in Parsons and Lear (1993) and in Parsons (1993). Historically the DFO management evolved from a primarily command and control system responsible for the research, assessment, allocation, licensing, regulation, and enforcement aspects of the fishery.

A fundamental element of fisheries management is the process of establishing and implementing regulations to maintain fish abundance at or above conservation targets. The regulations vary from setting quotas, Total Allowable Catches (TACs), and effort restrictions (e.g., minimum mesh sizes, maximum vessel sizes, season limits), through to regulations related to fish habitat manipulations (e.g., setting minimum water flow in salmon rivers). Despite the fact that the Fisheries Act contains apparently very strong powers to regulate industrial discharges or habitat destruction directly deleterious to fish, application of these provisions has generated considerable federal–provincial controversy, particularly on the west coast. However, there are rulings (e.g., Regina v. Sparrow 1991) that conservation responsibilities are the over-riding priority. Thus the goal of sustainability is legally entrenched in fisheries management in Canada. Indeed, the federal Department of Fisheries and Oceans has embraced and affirmed that goal in its own mission and operational mandates (see DFO 1994, 1995*c*, 1996*d*, 1997*b*). Recent debate on both the east and the west coasts has centered around the question of how far action may be from intent. The goals of conservation and sustainability are not particularly new. Indeed the key question is how to achieve these goals when social and economic goals, both long- and short-term, are seen to be at odds with the conservation goals.

Prior to 1993, scientific advice for Atlantic Canada fish stocks was provided by the Canadian Atlantic Fisheries Scientific Advisory Committee (CAFSAC) and its various subcommittees primarily composed of DFO biologists and research scientists (Hutchings et al. 1997; Doubleday et al. 1997). Subcommittee reports that summarized stock status and scientific advice were examined in committee and were not publicly available. The CAFSAC Steering Committee then prepared Advisory Documents summarizing the scientific information for the Atlantic Groundfish Advisory Committee (AGAC), consisting of industry and government (federal and provincial) representatives. This body provided recommendations to senior DFO management and ultimately the Minister. How final management decisions are made at the ministerial level is unclear.

As with many aspects of the Canadian marine fishery, the management system and decision-making processes have been recently undergoing significant change, primarily through the dissolution of CAFSAC in Atlantic Canada and through changes in the nature and diversity of advisory councils (e.g., the FRCC) and consultative committees, and briefing invitations to various user groups representing different sectors of the fishery. The proposed revised Fisheries Act (Bill C-62) may permit legally binding agreements or partnerships with other agencies or organizations (DFO 1996*b*). For example, on the west coast a number of partnerships have been developed with groups such as the Underwater Harvesters Association for the management of the geoduck fishery.

On the west coast, the division of responsibilities has now been clarified following federal–provincial negotiation under the 1996 Memorandum of Understanding (MOU) between Canada and British Columbia on Fisheries Issues (July 15, 1996). Responsibility for management of harvesting and habitat protection will be shared in various ways with the provincial government in British Columbia. The Canada–British Columbia Agreement on the Management of Pacific Salmon Fishery Issues (Department of Fisheries and Oceans 1997*c*) mandates creation of a Canada–BC Council of Fisheries Ministers, to meet twice per year, as well as formation of a federal–provincial Fisheries Renewal Advisory Board and a federal–provincial Pacific Fisheries Resource Conservation Council, all largely as recommended in the 1982 Royal Commission report (Pearse 1982) and again in the Fryer Report (1996). Development of a new Habitat Protection and Fisheries Enforcement Agreement is also envisaged, presumably to be linked with BC's newly introduced Fish Protection Act (1997).

In principle, and in general, there appears to be widespread scientific, public, judicial, and operational governmental agreement on a precautionary and risk-averse approach to the management of fisheries in the face of uncertainty. All participants have agreed, in principle, with the importance of maintaining harvests within sustainable levels. However, such rhetorical agreements acquire meaning only when founded upon concrete definitions of responsibility and operational action. Such definitive action presents a mighty challenge when the participants are faced with conflict over the interpretation of evidence reflecting the state of the resource and when faced with conflict over the policies to be adopted in regulating access and effort. It is here that apparent agreements often fall apart because the existing structures and institutional arrangements provide neither adequate consensus around decisions, nor adequate and explicit reflection of the uncertainties associated with decisions.

Two of the key problems in the present management arrangements can then be identified as a focus on harvest rather than on ecosystem health and a focus on present profit rather than on future stocks and future economic returns. There exist highly overcapitalized industrial fisheries, high geographic mobility of industrial fleets, global mobility of investment capital, and excessive community dependence on stocks for which adequate definitions of use or management rights do not exist. Each of these factors encourage a "race for the fish" and a built-in bias towards accepting the risks of overexploitation rather than accepting the current costs of possibly excessive caution with a view to future reward. Thus the system contains a structural bias against precaution and it encourages an excessive risk for resource collapse and possibly species extinction. Regulatory decisions intended to control harvesting pressures have a similar "built in" bias, for perfectly understandable organizational reasons (see Walters 1995; Munro and Neher 1997; Doern and Conway 1994), and that is the key point to be appreciated in describing or assessing the present state of the management system. The biases inherent in current structures are not signals of malevolence. They are the consequences of failures in development of a conservation ethic and in information flow and incentive systems, which lead to a situation in which rational individual action in the short-term overwhelms any view to a longer term cooperative outcome in which all can be better off. A simple example helps clarify this point. The price and interest rate signals provided by financial markets do not accurately convey community attitudes toward the risks of commercial extinction, or worse, species extinction. The social discount rates appropriate to continuing communities are normally dramatically lower than those of individual private owners or corporations operating on global capital markets. The decisions made by the managers of industrial fleets in weighing investment options on a global scale, when the goal is a maximum return to the owners of corporate shares, do not reflect the preferences of the local community residents who are among the public owners of the resource itself.

Box 2.9. Fisheries Management Goals and Biological Reference Points

A management program can only be evaluated in the context of its specific objective, the goals necessary to reach the objective, and the tasks undertaken to meet the goals. Management "reference points" provide concrete statements of the objective (Caddy and Mahon 1993). One example is target escapements for salmon spawners. A given target escapement is intended to produce a particular sustainable yield. In other cases, several distinct biological and economic reference points have been used and they are often based, at least in part, on "yield per recruit" analysis. For instance, there is "F_{max}," the harvest rate (or fishing mortality rate-related to the proportion of fish harvested) that results in the maximum biomass yield per recruit. A more conservative reference point frequently employed in Canadian fisheries management as a "target" reference point toward which management aims is the fishing mortality rate at which the rate of change in yield as a function of fishing mortality is 10% that of the rate of change of yield at very low fishing mortality rates. This reference point is designated "$F_{0.1}$" (Gulland and Boerema 1973). The $F_{0.1}$ strategy is an arbitrary choice, but it is used because it is a more conservative harvest rate than F_{max}, which is associated with a higher probability of stock depletion due to unforeseen factors such as environmental changes or increased harvesting efficiency per boat.

Other examples of reference points include limits on levels of exploitation that we do not wish to exceed, such as the fishing mortality rate that results in a reduction in spawning biomass per recruit to a specified percentage of the unfished population. For instance, $F_{20\%}$ would correspond to the fishing mortality rate at which the estimated spawning biomass per recruit was 20% of that at the unfished level. Such "limiting" reference points are typically used to reflect conditions that are to be avoided; many different reference points are used (Maguire and Mace 1993).

Limiting and target reference points have been routinely used in the management of Canadian marine fisheries since the inception of extended jurisdiction. For example, $F_{0.1}$ has been employed as a target reference point in management of groundfish resources in eastern Canada (Rivard and Maguire 1993; Maguire and Mace 1993) and western Canada (Leaman 1993), although for many fisheries the "$F_{0.1}$" target level was never achieved (see Box 2.4 for a prime example). Management of herring populations in western Canada is keyed to a target fraction (20%) of the preseason stock biomass estimate but is also subject to a minimum biomass constraint or cutoff level.

Conclusions

Many fisheries are now in trouble on both the east and west coasts of Canada and there is evidence that worldwide landings have leveled off with little likelihood of any further increase. Aquaculture is growing in importance but wild fisheries are still much more important and even with significant increases in growth aquaculture is unlikely to replace the wild fisheries. No simple solution to the fisheries problems lies in sight although we are gaining greater insight into the problems. The natural and social sciences offer differing perspectives on our present situation on questions of overcapacity, management, habitat protection, and environmental fluctuations. With our present knowledge what should we expect in the future in the face of inevitable change? In the following chapters of this report we discuss in some specific detail the processes of change and their related uncertainties from the climatological, environmental, biological, sociological, or economic perspectives as they pertain to Canadian fisheries. We also examine the risks created by these uncertainties and develop principles and strategies for responding to them.

Chapter 3. The Nature of Change

Introduction

The challenge for fishery managers, the fishing industry, fishermen, and coastal communities is to maintain marine ecosystems with biologically sustainable fish populations that produce substantial ongoing economic and social benefits. These goals are especially difficult to achieve in the presence of the numerous types of change that inevitably occur in fishery systems. This chapter reviews changes and places them in a broad context by first considering how fish change abundance varies. The chapter also reviews specific examples of how fishery systems are affected by changes in the physical and biological environment, economic processes, and socio-political factors.

To begin, consider the components of change in abundance of a fish stock (usually measured by total biomass of the stock). Variation in stock biomass from one year to the next results from a combination of gains and losses (Fig. 3.1). Gains arise from growth in body size of fish that survive during that year and from reproduction (recruitment of offspring into the population). Losses occur from those fish that die from natural causes and those that are harvested.

Human economic and social benefits mainly derive from the harvesting component of Fig. 3.1. As described in Chapter 2, Box 2.2 Sustainable Harvests, because of the biological responses of fish populations, it is usually possible to harvest some biomass annually. However, to obtain a fish stock that is sustainable in the long-term, the combined losses from natural mortality and harvesting must *not* repeatedly exceed the gains from growth and recruitment, otherwise the stock biomass will decrease toward zero. The task of adjusting harvest levels to meet this condition might be relatively straightforward if (*a*) all components

Fig. 3.1. Components of change from year to year in biomass of a fish stock, illustrating those components that tend to increase biomass (upward arrows) and those that tend to decrease it (downward arrows).

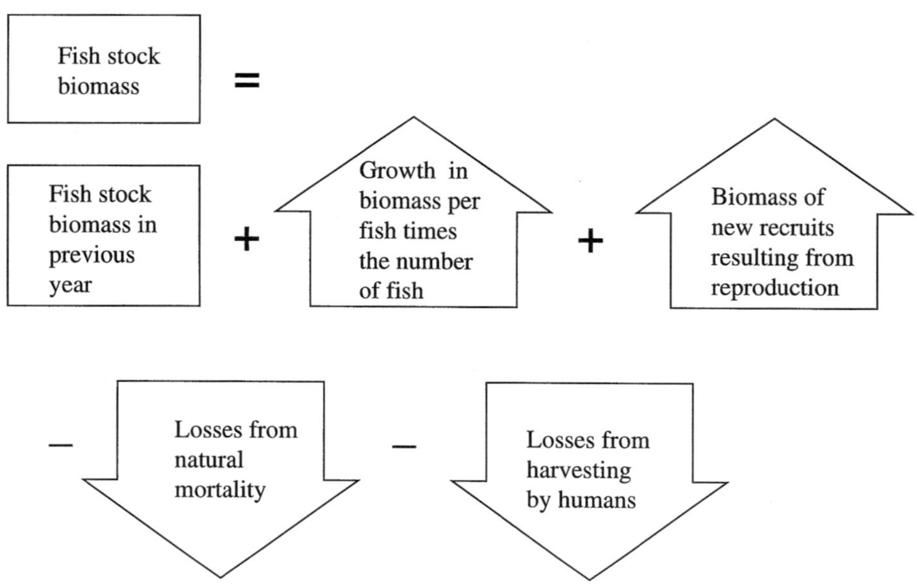

of Fig. 3.1 were known with certainty; (*b*) these components were either constant over time and space or, if variable, then at least accurately predictable in a timely manner; and (*c*) harvests were perfectly controllable. However, *none* of the above conditions is true. All components of change in fish stock biomass are variable and uncertain. They are not constant over either time or space, nor are they accurately predictable. As well, it is not possible to take exactly some specified harvest due to the uncertain dynamics of a fishing fleet.

One reason for this lack of certainty, predictability, and control is change — change in all components of fishery systems. These systems are dynamic because the fish and the humans involved are affected by constantly varying physical, economic, socio-political, and biological processes. For instance, changing physical oceanographic conditions can drive variability in growth and survival rates of fish, as can the abundance of fish. In addition, socio-political and economic processes cause variability in prices, changes in costs of fishing, increased harvesting capacity, changes in spatial distribution of fishing vessels, and increased technological sophistication of fishing gear. These sources of change also have complex feedback linkages with one another (Fig. 3.2), which further add to uncertainty, unpredictability, and lack of control. An economic "balance sheet" for fishermen would be composed of components analogous to those in Fig. 3.1, with gains related to the value of fish caught and losses related to costs of investments in fishing gear or costs of monitoring.

Fig. 3.2. First-order, or direct, interactions among major physical, biological, economic, and socio-political components of fisheries systems, focusing on their effects on changes in fish stock abundance and amount harvested. Differently shaped boxes indicate different types of components: economic (non-boldface rounded-corner boxes), biological (boldface boxes), socio-political (parallelograms), and physical (ellipses).

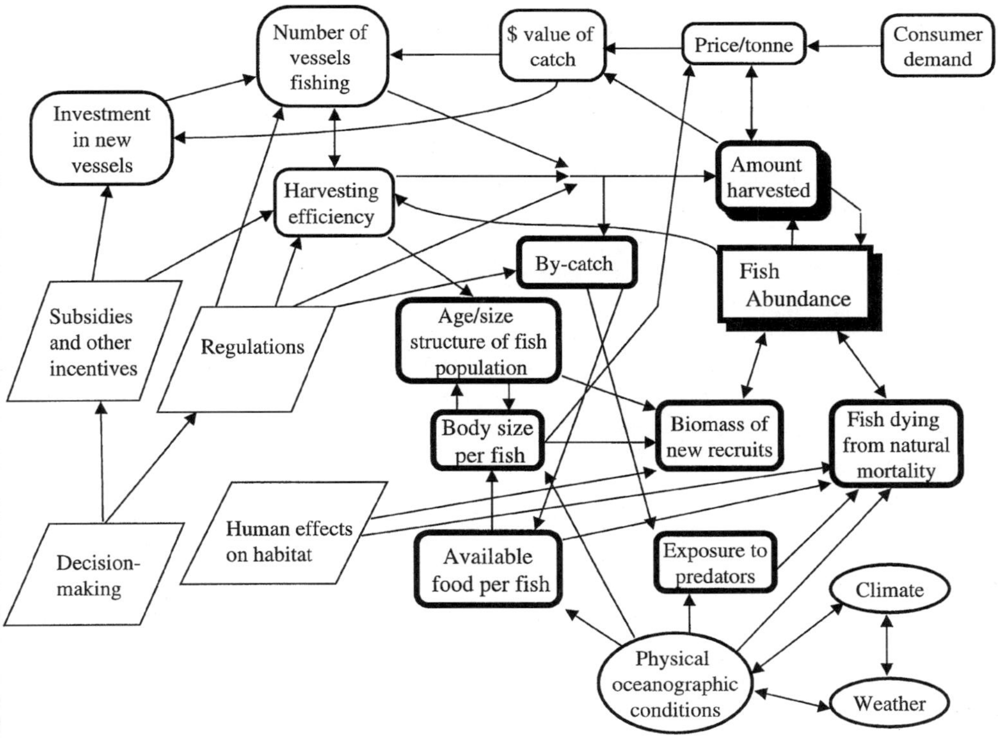

As we show below, these components are as changeable and unpredictable as the biological ones.

Because of these sources of change, a constant annual amount of biological and economic yield is not likely to exist. Instead, year-to-year variability and more persistent changes are the norm and those involved in fisheries, either as resource users or resource managers, must expect this. The challenge is to adjust appropriately the factors that are under human control (e.g., harvesting and habitat disturbance, which influence recruitment) in order to ensure that losses in fish biomass from harvesting and natural mortality do not repeatedly exceed the gains due to growth and reproduction.

To understand how to meet this goal, this chapter describes both natural and human-caused mechanisms of change and emphasizes how they create uncertainties. Chapter 4 then discusses how these uncertainties create risks for fish stocks and for people who depend on them. These ideas will provide the foundation for principles (in Chapter 5) and feasible strategies (Chapter 6) for achieving sustainable fisheries in the presence of unavoidable variability and uncertainty. These principles and strategies should be followed by the fishing industry, fishing-dependent communities, fisheries management agencies, and society in general

We first describe the variability and change in some of the biological components shown in Fig. 3.1. We then discuss physical environmental processes, followed by economic, socio-political, and biological processes. Readers should keep in mind that, while the discussion here is necessarily structured sequentially, complex interactions occur among these different types of processes (Fig. 3.2). For this reason, biological processes are described in two sections because, in order to properly interpret data on them, we need to examine them in the broader context of other factors that have changed at the same time, such as size-selective harvest rate, productivity of food for fish due to oceanographic variation, etc.

Biological variability and change — a brief overview

To put the rest of this chapter in context, we examine the variability and change that has occurred in the three biological components of fish population dynamics shown in Fig. 3.1: mortality rate, growth rate (in biomass per fish), and reproductive rate. These components significantly affect the magnitude of sustainable harvests by humans, so two important questions are, how variable are these components and how persistent have changes been?

Survival rate of fish between the egg stage and recruitment to a specified age is typically highly variable across time within a stock. In Pacific salmon, for instance, many biological and environmental processes affect survival rate. A cohort of eggs and the resulting juveniles are influenced by predation, competition for food, and density-independent processes such as freshwater flow rates and upwelling caused by winds. Variation in these factors produces a wide range of recruits per spawner from year to year, as is illustrated in Fig. 3.3 by four frequency distributions typical of B.C. salmon stocks from data series over 40 years long. Another feature of the recruits per spawner estimate is that such data tend to be autocorrelated, with lengthy periods having above- or below-average recruits per spawner (Walters 1987). Such autocorrelated patterns often suggest hypotheses about environmental mechanisms causing changes (Mann and Lazier 1996).

In the case of herring, flatfish, and other such species where eggs hatch in the ocean as extremely small larvae, survival rate is also affected by factors such as ocean currents that can displace larvae to inhospitable habitats (Sinclair 1988). In such marine fish species, it is

Fig. 3.3. Frequency distributions of recruits per spawner for four Fraser River, British Columbia sockeye salmon stocks over brood years (year of spawning) 1949–1990.

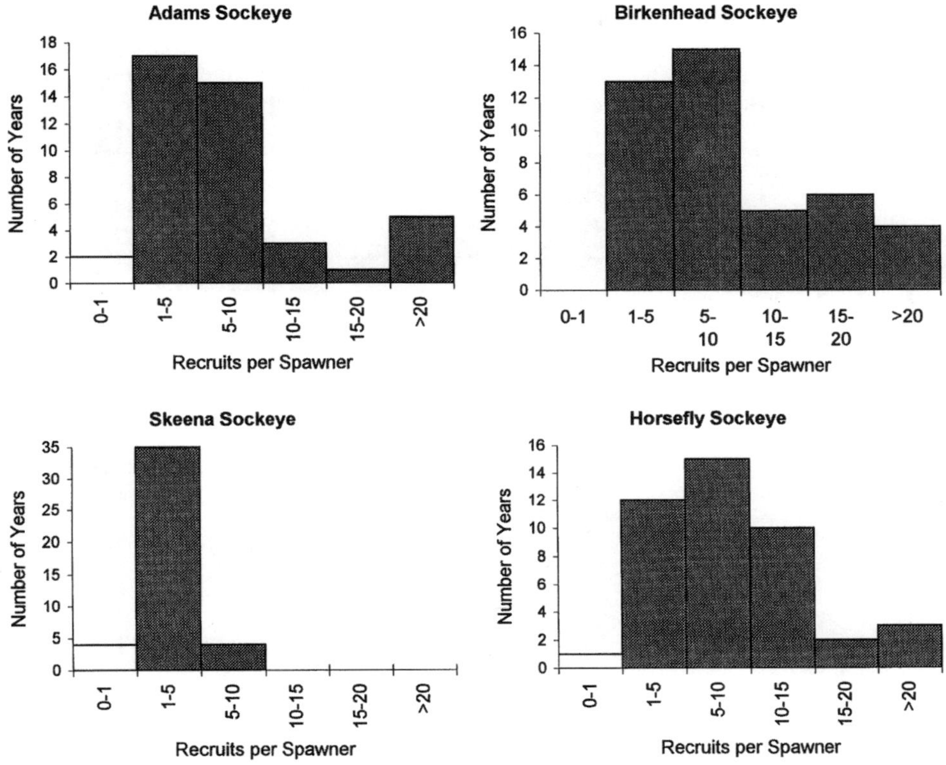

common to see a wide range of variation in recruitment to some age (Fig. 3.4).

Although growth rate of fish body size tends to have less variation than survival rate, it is an important source of change for several reasons. First, the price per kilogram tends to be higher for large fish (e.g., Fig. 3.5 for lobster) due to there being less waste per tonne during processing. Second, a given number of small fish will obviously produce lower total tonnage than large fish. Fig. 3.6 shows that not only is the average body length of 4-year-old Fraser River adult sockeye salmon variable from year to year, but the average size has decreased about 5% over the entire period, with most of the decrease occurring since the early 1980s. Thus, Fig. 3.6 shows both variability and change — year-to-year variability in body size has been superimposed over a longer term, more persistent downward trend in mean size. Other studies (e.g., Ishida et al. 1993; Bigler et al. 1996) also show a significant decrease through time in body size of North Pacific salmon, although most of the data in those papers were not age-specific body size and hence were potentially confounded with changes in age at maturity (and the resulting vulnerability to fishing).

In addition, body size is important because larger females tend to produce disproportionately more eggs than small females. Hence, if a population of long-lived fish becomes on average younger and smaller due to fishing, the total reproductive output from the population could be reduced substantially (unless age at maturity decreased sufficiently to compensate, or egg size and survival rate increased enough to compensate). This decreased

Fig. 3.4. Frequency distribution of year-classes of recruitment for 9 different marine fish stocks. The X axis (recruitment interval) indicates how large the annual recruitment was relative to the mean recruitment for each stock (after Hennemuth et al. 1980). Units are in terms of the stock's mean standardized recruitment divided by 5 (i.e., a recruitment interval of 5 equals the mean recruitment; a value of 10 on that X axis is twice the mean recruitment). For instance, for Norwegian spring-spawning herring (bottom), one year (at the recruitment interval of 10) had a recruitment twice the long-term mean recruitment.

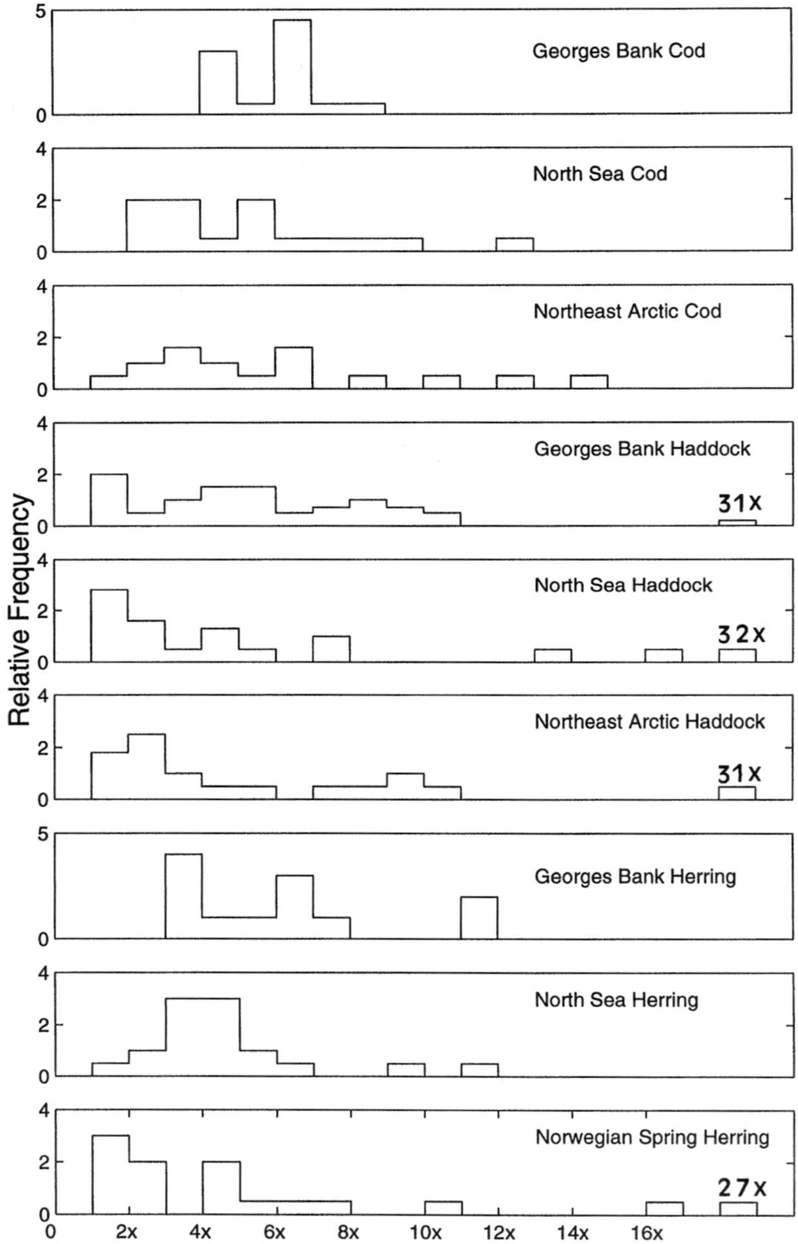

Fig. 3.5. New England wholesale lobster prices for different grades (size-classes) for the period 1987–1990 (after Pringle and Burke 1993).

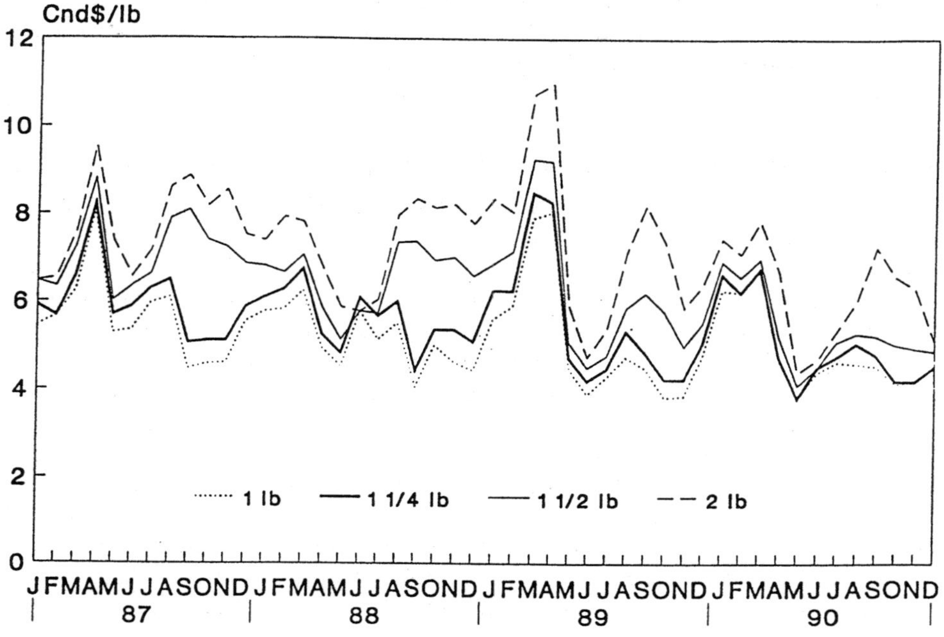

Fig. 3.6. Average standardized residuals in body length of 4-year-old Fraser River sockeye salmon, averaged across 21 stocks. Data for each stock were standardized to their respective mean sizes before being combined into the index shown.

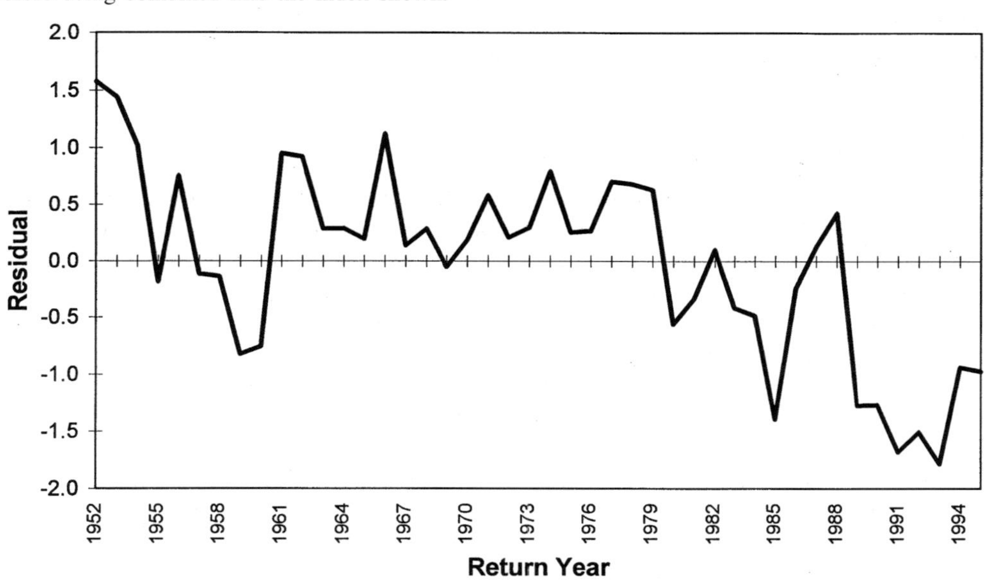

reproductive output would reduce the ability of the population to cope with adverse environmental conditions. Such a change in the size structure toward younger, smaller fish probably contributed to the recent collapse of cod in Atlantic Canada (Hutchings and Myers 1994).

This chapter attempts to describe the interactions of these biological processes, as well as physical oceanographic, economic, and socio-political processes, in order to understand the nature of variability and change in fishery systems.

Physical environmental processes that affect fishery systems

Physical environmental processes affect fish through various mechanisms. Processes in the atmosphere and the ocean occur over time scales ranging from short (daily or less) to long (hundreds to thousands of years, as with climate change). This report focuses on four time scales. (1) *Long-term processes* — changes that occur with a very long time scale, or that persist for many decades or centuries (e.g., ice ages, anthropogenically caused global climate change). (2) *Regime shifts or decadal-scale processes* — changes of the ocean environment that occur over, or persist for, periods up to a few decades. When the changes appear abruptly, they are sometimes referred to as regime shifts. Such regimes may persist for one to two decades and may involve ocean currents, temperature, ocean productivity, and other environmental variables. The shift in the North Pacific that occurred in the mid-1970s (discussed later) is an example of a decadal-scale process. (3) *Interannual processes* — changes that last from a few years to a decade; included in this category are quasi-periodic events (those that occur at irregular frequencies) such as the El Nino-Southern Oscillation (ENSO) and the North Atlantic Oscillation (NAO). (4) *Short-term processes* — changes that occur over periods ranging from daily, e.g., tidal, to seasonal, and annual, e.g., changes in wind-driven upwelling (which brings nutrient-rich water up to the surface) and downwelling (which pushes surface waters down).

The distinction among these time scales is somewhat arbitrary and many environmental processes cover more than one scale. Interestingly, the economic, socio-political, and biological processes that follow later in this chapter also create changes at various time scales and for discussion purposes they may also fit usefully into these general categories. Furthermore, these widely different sources of change create uncertainties at these time scales, which help to identify general principles for dealing with them.

1. Long-term physical processes

Change in the natural system can be caused by long-term trends in climate and the marine environment caused by human activity and those resulting from natural fluctuations. However, it is usually difficult to distinguish between these causes when examining their effects in the marine environment and in fish populations. This is partly because, by definition, long-term physical processes change slowly or persist in some state over long periods (e.g., centuries) and few such long records exist for the ocean environment. Most scientific environmental measurements were started in the last century in the northern hemisphere, although time series can be extended further back in time using proxy records such as the isotopic ratio of elements found in glacial ice cores to estimate the air temperature for the past 900 000 years (Fig. 3.7). This record shows that variability and change in global temperature is quite common. We are presently in a relatively warm period, quite unusual in the climate record. Temperatures over the past ten thousand years have been warmer than at any time in the previous ninety thousand years. Figure 3.7 also shows that the range of temperature variability increases as one goes back in time covering several degrees Celsius.

Fig. 3.7. Temperature over the past 900 000 years as derived from ice cores. Note the different time and temperature scales in each of the three panels.

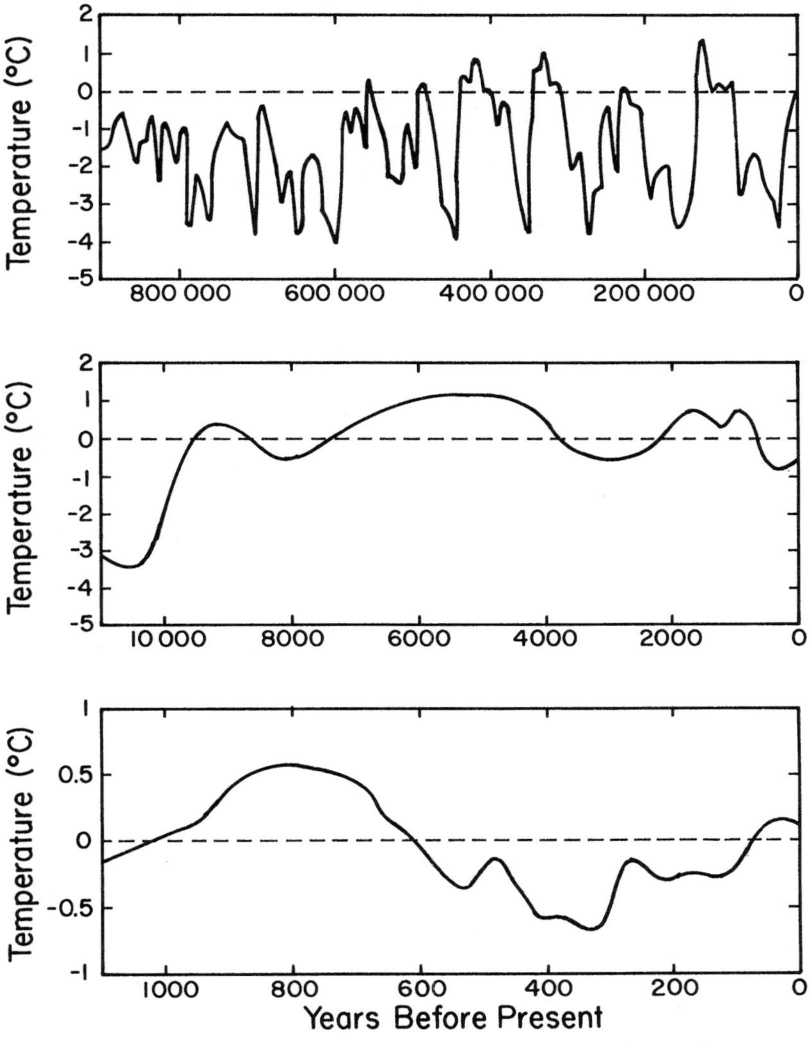

The connection between these atmospheric changes and the ocean is very strong. At these long time-scales, it is the slow response of the ocean that controls the climate system because of the ocean's great heat capacity. Just a few metres of water have the same heat capacity as the total atmosphere above it. At these long time scales, it is the large, basin-scale currents that are important. The exchange of heat with surface waters and the consequent formation of deep water result in the flushing of deep water to the surface. The time to mix the whole ocean is hundreds of years.

The possibility of a role of human influence on long-term climate change is often discussed and is the subject of intense debate. Since the late 19th century, carbon dioxide levels have increased by almost 25% with forecasts that by some time in the next century they will have doubled from their pre-industrial levels. The effects of other increasing industrial greenhouse gases, such as methane and nitrogen dioxide, must also be considered. The recent review of climate change by the Intergovernmental Panel on Climate Change (IPCC 1996) concluded that, through the increased production of greenhouse gases, humans had contributed significantly to the recent observed warming (Fig. 3.8). The model predictions for the coming century suggest that sea level will continue to rise, perhaps by as much as another metre, and that global mean temperature will rise by between 1 and 3.5°C. The maximum temperature changes are projected to take place over land in the northern part of the northern hemisphere, i.e., Canada and Russia. The projected climatic changes are expected to lead to changes in sea surface temperature, sea-ice distributions, surface winds, and ocean currents. There is as yet little consensus on the details of the oceanic response, but there is growing acceptance of a human-induced change, and it is expected that oceanic warming will be greatest at higher latitudes in the North Atlantic and Pacific (IPCC 1996). The northwestern North Atlantic is expected, however, to remain cool for the next few decades. Such changes could lead to distributional shifts in species that are sensitive to temperature, such as cod and salmon.

Most scientists now accept that global warming will occur — debates mainly concern how, when, and where, not "if." Over the next few years and decades, this uncertainty of the effect of greenhouse gases will decrease as our models and understanding of the climate system improve and the climate record is lengthened and broadened. The levels of carbon dioxide and other greenhouse gases will continue to increase for the next several decades; changes in human behavior and consequent reductions in emissions of the gases will not be

Fig. 3.8. Combined land, air, and sea-surface temperature anomalies (1861–1989) relative to the 1961–1990 mean for the entire globe.

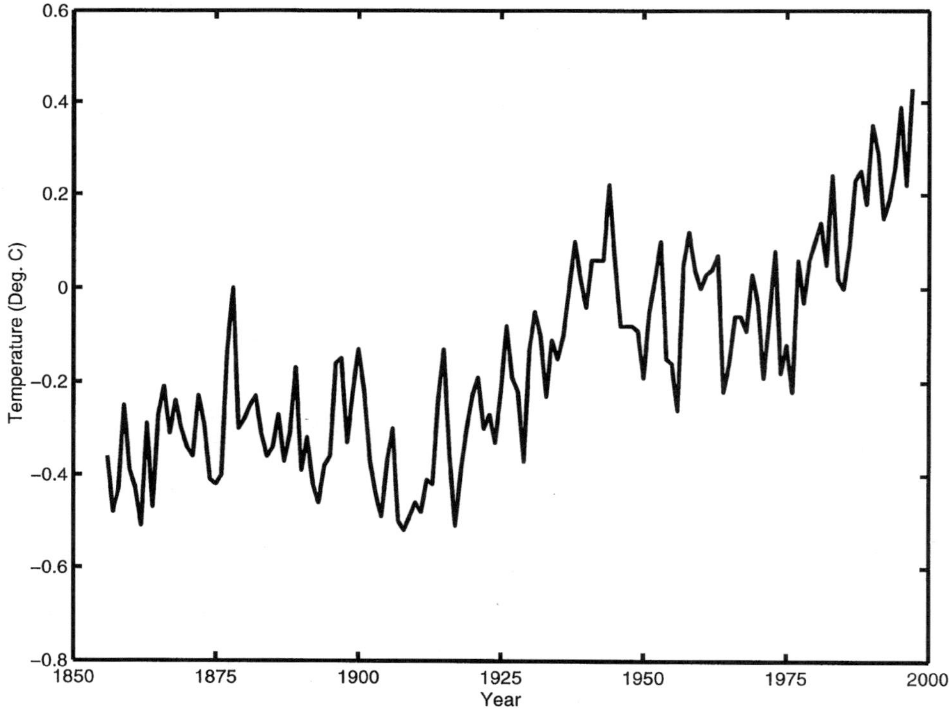

reflected for some time.

Early studies of global warming due to greenhouse gases, going back to the mid-nineteenth century, assumed that climate changes would be fairly gradual. Recently, improved analysis of ice cores and sophisticated coupled ocean–atmosphere models have suggested that sudden changes in the cooling of polar surface waters, downward movement of surface waters, and the formation of deep water may quickly and dramatically change climate over a few decades (Dansgaard et al. 1993; Manabe and Stouffer 1995). One useful conceptual model of the ocean considers the deep water movement as a large conveyor-belt system (Broecker et al. 1985) in which cold deep waters move from the northern North Atlantic regions to the Antarctic and then around the global ocean while warming and rising to the surface, thus providing a return flow to the regions where cooling and sinking occur. When first developed (Stommel 1957), this model was thought to imply slow and steady changes, but newer results suggest that even at the basin scale, this ocean system may change over as little as a decade. Significantly, after such an abrupt change occurs, the new condition may persist for a long period, potentially leading not only to significant biological changes, but also major economic and social changes if mechanisms for humans to adapt quickly are not in place.

Another long-term process that is potentially relevant to Canadian marine fisheries is atmospheric ozone, which is being depleted primarily due to the releases of human-produced chlorofluorocarbons (Solomon 1990). Changes in the concentration of ozone have direct effects on the levels of UV-B radiation at the earth's surface. A growing body of literature shows that this radiation detrimentally affects the physiology and growth of a range of marine organisms (Voytek 1990), including phytoplankton, the primary producers in the ocean, but also secondary producers like zooplankton. Organisms at higher trophic levels can also be harmed. For example, Malloy et al. (1997) show that eggs of Antarctic icefish exposed to ozone-induced UV-B radiation exhibit DNA damage identical to that observed in micro-organisms. The ecosystem effects of this increase in UV-B radiation are as yet unknown.

2. Regime shifts in physical processes — decadal-scale variability

Over the past thirty years detailed measurements have been obtained of many biological and physical systems in the ocean. These time series have revealed that decadal-scale change is a fundamental characteristic of oceans and that all ocean ecosystems have complex biological and physical interactions on these time scales. Changes in the physical environment can influence biological components independently. For instance, fish growth, primary production, and interactions among other biological components occur at the various spatial and temporal scales of the physical environment. This means that a single stable ocean ecosystem, about which some small perturbations occur, probably only exists in a few isolated circumstances. We should expect a myriad of possible states and changes.

In the North Pacific, for example, a relatively abrupt change was observed in the strength of the North Pacific Index, a measure of atmospheric and oceanographic conditions in the Gulf of Alaska. This index had at least two shifts over the last 100 years; one occurred in the mid-1970s (Trenberth and Hurrell 1995; Fig. 3.9). This was associated with changes in wind speed, patterns in ocean currents and upwelling, and other oceanographic processes, which in turn apparently had large effects on salmon stocks in the North Pacific (Hare and Francis 1995; Peterman et al. 1998).

Fig. 3.9. Time series of the North Pacific Index (NPI), which is the mean North Pacific sea level pressures averaged over 30°N to 65°N and 160°E to 140W) for November through March beginning in 1925. Data were smoothed with a low-pass filter (slowly curving line) (after Trenberth and Hurrell 1995).

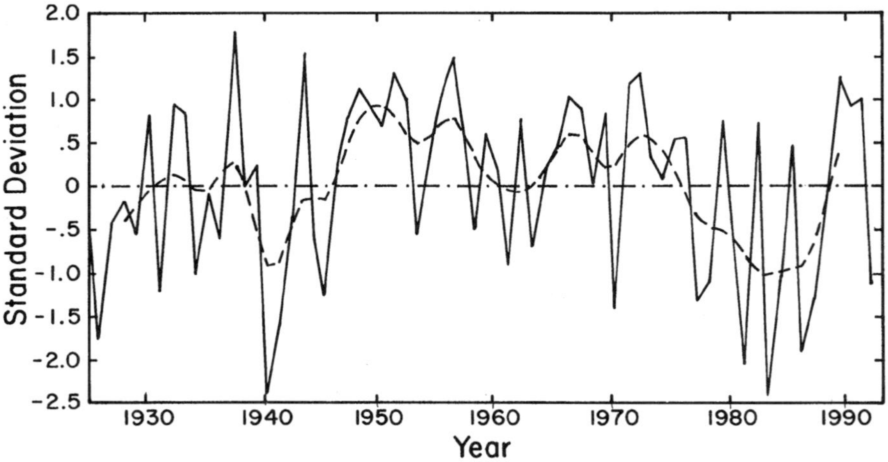

Sea surface temperature in the North Atlantic shows clear shifts in some areas over the past decades with cool periods (1910 – late 1920s, 1970 – late 1980s) and warm ones (1940–1960). There is also a strong correlation between air temperature and sea surface temperature, supporting the use of air temperature as an index of sea surface temperature. These observed changes can be related to observations elsewhere on the globe (Fig. 3.8) and locally within the North Atlantic (Fig. 2.8). However, changes in the northwest Atlantic are not always synchronous with changes elsewhere in the Atlantic. For example, temperature increases in the northwest Atlantic are correlated with decreases in temperature across the basin in the northeast Atlantic (Dickson and Brander 1993).

3. Interannual physical processes and environmental events

Long-term and decadal-scale changes are not the only time scales of change. Significant environmental variability also occurs between annual and decadal time scales. In the North Pacific, much of this interannual time scale variability is associated with El Nino – Southern Oscillation (ENSO) events, in which warm waters move along the equator to the eastern Pacific and subsequently northwards up the west coast of North America. Although there is much that we do not know, it appears that the observed ENSO-related variability in the equatorial Pacific leads to direct but lagged effects throughout the Pacific, including effects on fish populations that are described later. The environmental effects of ENSO are considered so important that the United States has deployed a permanent oceanographic array of instruments along the equatorial Pacific to provide data for monitoring and forecasting ENSO events.

Interannual variability also exists in the North Atlantic, much of it being associated with the North Atlantic Oscillation (NAO) (Dickson and Brander 1993). This oscillation involves air pressure differences from near the equator to the northern Atlantic. A strong pressure difference leads to greater-than-average northwesterly winds and cooling conditions in the northwest Atlantic, whereas a weaker pressure difference results in reduced northwesterlies and warming conditions there. Recent oceanographic studies of water masses have led to the suggestion that ocean currents of the North Atlantic may act as the "spring" to determine the periodicity of the NAO "clock" (McCartney et al. 1997).

4. Short-term physical events

The effect of short-term physical environmental changes on fishable biomass depends upon the life stage of the fish affected and the space–time scales of the change. In general, the shorter the time scale, the smaller the spatial scale, but there are exceptions. Wind-driven turbulent effects, for example, have very small spatial and short temporal scales (centimetres and seconds) but cover an area associated with the larger scales of wind patterns. Thus, seasonal or annual changes in winds at the ocean basin scale can influence large-scale oceanic mixing, and changes in wind forcing associated with climate change or a regime shift can lead to changes in short-time scale ocean events.

How these short-term changes in the physical environment influence fish is still a topic of active research (Bakun 1997). The greatest effect is most likely for fish larvae, since the larvae are small and therefore more susceptible to changes in the ocean environment, food supply, and predation. Recent oceanographic studies in the Atlantic (US GLOBEC program on Georges Bank — Wiebe and Beardsley 1996) and in the Pacific (FOCI in Shelikov Strait — Kendall et al. 1996) have shown how the currents can influence the survival and success of larvae of different species. These studies show that with intensive effort we can now begin to understand some of the links between the physical environment and fish growth and survival.

Over the past few decades, harmful algal blooms, sometimes called "red tides," have increased in frequency, severity, and duration (Anderson 1989; Smayda 1990). While in the past such blooms were isolated to a few regions, they are now widespread. In many cases, the blooms extend over large geographic areas and are composed of more than one toxic species. They can cause human death from paralytic shellfish poisoning. They also have an economic impact due to costs arising from closures of shellfish fisheries, mortalities of wild and farmed fish, and monitoring programs for the toxins. The causes of the observed increased occurrences of blooms are unknown, but it is believed that human-influenced degradation of water quality in the coastal zone is an important factor.

We generally understand short-term events better than long-term ones because we have more observations of the former. Thus, tidal processes and changes in light level in the ocean are quite well-understood. Light level is primarily affected by latitude, but it can change very quickly since it also depends upon cloud cover. Thus, weather can influence light levels, which can subsequently control timing of the spring production of plankton (a primary source of food for many fish larvae) and thereby influence fish productivity (Mann and Lazier 1996). At the seasonal time scale, changes in the production and melting of sea-ice on the Labrador coast determine the surface salinity and thus determine the stratification of local marine waters. This has the potential to change food production for fish larvae during the spring and summer. Such changes in the seasonal and daily cycles can be quite abrupt. Specific examples of the biological effects of short-term physical processes and their hypothesized mechanisms are discussed in a later section.

In short, these physical oceanographic processes that are driven by climate and weather create variation at several temporal and spatial scales. This temporal variation affects fish populations and thereby creates uncertainties about their future productivity and harvest.

Economic processes that affect fishery systems

Within the complex array of economic, social, and political institutions, a wide variety of economic processes directly influence the human and natural components of fishery systems. These include fluctuations in prices, changes in demand, globalization of markets, development of new technologies in the fishing industry, and the dynamics of capital investment. Collectively, these processes lead to uncertainty for fishermen, industry, communities, and management agencies.

1. Variations in price

The economic sustainability of fisheries can obviously be affected by variations in prices paid to harvesters and processors. Prices for fish often fluctuate widely over time, based on both their availability (e.g., whether stocks are abundant or whether the fishing season is open or closed in a particular area) and on the availability of substitutes (e.g., other fish species or other foods). Such changes can occur over a few days, over the course of a fishing season, or from one year to the next. For example, wholesale prices for lobster sold in the New England market dropped by about 45% over one month in 1989 (Fig. 3.5; Pringle and Burke 1993).

The effect of these fluctuations on economic viability of fishing operations is thus similar to the effect of short-term environmental fluctuations on the survival rate of fish; both types of fluctuations can be absorbed as long as the period of unfavorable conditions does not persist too long. In the case of the fishing industry, the definition of "too long" depends on the specific situation and alternative sources of fish and revenue.

In addition, variability in prices can influence the behavior of vessel owners because their choices about going fishing depend in part on prevailing prices, particularly at the start of a season. For instance, higher prices will lead more vessels to go fishing than might normally be expected, all else being equal. Conversely, lower prices may result in lower participation rates. Thus, fluctuating prices can lead to inaccurate forecasts of total fishing effort during a given fishery opening, which can lead to management decisions that increase the probability of undesirable harvest rates. Extreme increases in prices may require managers to impose severe restrictions to prevent overharvesting. For example, high prices for B.C. herring roe in 1979 induced many purse seine vessels with high catching power to try to participate in the fishery. However, with the openings already as short as 15 minutes per year in some areas (only enough time for one set of the nets), managers were forced to severely restrict the number of vessels allowed to fish to reduce the probability of overharvesting.

Not only do changes in prices of fish affect decisions about the proportion of fish harvesting capacity that is utilized within a year, but they also influence decisions about creating new harvesting capacity. For example, the relatively high prices for some species in recent decades has provided a strong incentive for new boat-building and improvement in fish-finding and harvesting technology, aided by government subsidies. The correlation between fish prices and boat-building is certainly imprecise, since the latter depends on other factors, notably fish abundance. Nevertheless, the connection was strong in the late 1980s for groundfish vessels under 100 feet in length in the Scotia-Fundy management region of Atlantic Canada. The rate of vessel construction within that fleet rose over the period 1984–1988 in conjunction with an increase in the average dockside price per pound for groundfish (cod, haddock, and pollock).

Another issue related to prices is the tendency for commercial harvesting to occur most heavily first on the most valuable and (or) most accessible species, then when those have been depleted, to shift onto the next most valuable and (or) accessible species. This occurred, for instance in global whale fisheries, which first targeted mainly large and valuable blue whales, then shifted onto less-valuable species — first fin whales, then sei whales, then even smaller minke whales (Clark and Lamberson 1982).

Uncertainty about prices and their effects tends of course to be particularly great in new and emerging fisheries where experience is limited or non-existent.

2. Changes in demand

Short-term price fluctuations are driven by many factors, notably the availability of the product and its substitutes. In the longer term, however, the fundamental matter from an economic perspective is that of demand — the relationship between the quantity of a product desired by consumers and the product's price. Demand usually tends to vary over longer time scales than the price level, responding to the ever-growing human population, changes in the spatial scale of the market (i.e., who has access to the product), shifting consumer preferences, or changes in average income levels.

However, shifts in demand can also occur rapidly, in the form of jumps ("regime shifts" in our temporal categories). For example, a rapid increase in demand for fish was driven in part by publicity over the idea that eating fish was healthier than eating red meat. This new regime was sustained over a substantial period, although there was some decrease in North America following publicity over possible contamination of fish from pollution. Such changes in consumer behavior can create economic uncertainties for industry and management and constrain their ability to make good long-term plans and decisions.

3. Globalization of markets

Throughout the economy, globalization of markets has been extremely important. Markets for many goods, as well as for currency and other financial instruments, have changed from being local, to national, and then international. Trade has increased dramatically on a global scale. Globalization of the financial system over the past several decades (and a corresponding widespread adoption of flexible exchange rates) has increased variability and uncertainty in exchange rates. It has also increased the extent of currency speculation.

While many fisheries in Canada have been based on international trade for centuries, the above interrelated changes have had major impacts on fisheries in recent years.

(*a*) The *geographical* attachments among consumers, producers, and processors of raw fish product have diminished greatly. A remarkable example of this comes from the recent collapse of the northern cod stock off Newfoundland; while a major event in Atlantic Canada, the collapse had little impact internationally on the fish market. This is because, especially since the late 1970s, a worldwide market emerged for "whitefish," which includes not only Atlantic cod but also North Sea haddock, Alaskan pollock (landings of which increased greatly as northern cod decreased), and many other species. There is now great substitutability among the various whitefish species, so that with the loss of any one contributor to the supply of whitefish (such as northern cod), wholesalers and retailers merely switch to another source of whitefish. This is very different from past fish stock collapses; fisheries on such stocks tended to serve specific markets, which were thus strongly affected by the collapse. Indeed, globalization of markets and substitutability among species imply that if and when the northern cod stock recovers to produce substantial annual catches, it will be a challenge to recover the "market niche" the stock used to have internationally.

(*b*) The attachment between Canadian fish processing companies and Canadian fish resources has also diminished. For example, integrated processors in Atlantic Canada, such as National Sea Products and Fishery Products International, which were formerly reliant on local fish stocks, have reorganized to process fish imported from anywhere in the world. This reorganization and a greater concentration on product distribution, have made these companies essentially independent of the local supply of fish. This decoupling between local supply and the output of fish processors may make processors more vulnerable to worldwide economic and environmental trends, but it also may make them less vulnerable to local fluctuations in abundance and thereby reduce their incentive to conserve the local resource. This expectation is based on observations elsewhere that long-term reliance on a fish resource, as found more with local fishermen and small-scale processors, provides a major incentive for proper stewardship of the resource (Berkes and Folke 1994; Pinkerton and Weinstein 1995).

(*c*) The increasing variability in exchange rates, given an export-based fishery, induces fluctuations in foreign demand for locally caught fish (as the fish becomes alternately more expensive or less expensive to buy in other countries). This increases the uncertainty in (*i*) financial returns to be obtained from any given harvest, (*ii*) the returns from a given investment in harvesting capacity or processing plants, and (*iii*) the benefits arising from "investing" in the natural capital — the healthy fish stock — to generate future revenue. In turn, this uncertainty may affect the behavior of fishermen and in particular the level of fishing effort chosen (especially if there are alternatives to fishing available). Notably, it can be anticipated that such uncertainty may induce fishermen and investors in such a fishery to take a shorter time horizon, leading to adoption of riskier and shorter-term harvesting practices.

Some impacts of globalization on fisheries are unclear due to simultaneously induced changes. For example, if globalization makes fishery systems more sensitive to external economic conditions, so that industry and managers are less able to accurately forecast prices and thereby control fishing effort, decision-making at the local or regional scale may be hampered. Alternatively, if fish prices happen to be more stable at the international level, globalization could reduce variability in fishermen's incomes.

Economic globalization can also interact with environmental factors to further complicate the balance of uncertainty and risk. For example, the lobster fishery along the Atlantic coast of Canada and the U.S. is relatively concentrated geographically and produces a unique product (i.e., Atlantic lobster is viewed as different from other shellfish, unlike the "whitefish" situation). Thus, the supply of lobster has a relatively large influence over international prices. If a widespread environmental change occurs over much of the lobster's geographical range, affecting abundance and harvests, this may tend to affect price. Indeed, if lobster harvests decline, so that less lobster is available on international markets, this will tend to increase the price (other things being equal), thus helping to stabilize incomes of fishermen and reduce risks.

Pacific salmon also exemplify the interaction between economic and environmental factors. As will be discussed later in this chapter, an apparent environmentally driven increase in productivity of food for salmon in the Gulf of Alaska in recent decades led to a large *increase* in abundance and catch of Alaskan salmon. This, in conjunction with the entry of Russian salmon into world markets (Pinkerton 1994) and a greater availability of salmon from aquaculture facilities, caused a large *decrease* in price per kilogram of salmon. The impact of this price decline might have been lessened through more innovative marketing. For instance, efforts to increase salmon markets (e.g., by the Alaska Seafood Marketing Council) could have focused less on the traditional salmon canning industry and more on absorbing the increased salmon abundance through development of new products and markets for them.

4. Technological development

Advances in fishing technology that led to greater efficiency in harvesting have long been a source of change in fisheries. This includes both gradual change that occurred over decades or centuries and occasional "regime shifts" in which major new fishing technologies were adopted widely and rapidly. Improved fishing technology often led to higher annual harvests, at least initially, and an increased probability of stock collapse if the greater harvesting capacity was not properly controlled.

Development of the northern cod fishery over the past few centuries illustrates the effects of technology on harvests (see Box 2.4 on northern cod). Hutchings and Myers (1994) documented that annual catches roughly doubled during the 19th century to about 200 000 tonnes per year after introduction of several technological advances in fishing methods. Then, in the mid-1950s, more efficient stern-hauled otter trawlers led to an even more dramatic increase in annual catches. While most of the catch prior to the mid-1970s was by foreign fleets, Canadian technology and catches quickly caught up. This rapid change in gear composition of the fleet might be viewed as a regime shift in technology to a higher level of harvesting efficiency. Unfortunately, in the 1970s, the stock collapsed, rebuilt somewhat in the early 1980s, but then collapsed again in the early 1990s.

Of course, environmental conditions and other economic factors such as consumer demand, size of markets, access to markets, and capitalization of the fleets undoubtedly played a role in affecting the northern cod stock (Box 2.4 on northern cod). However, it has been argued persuasively (Cadigan et al. 1993; Hutchings and Myers 1994 and references therein) that change in the technological capability and consequent increase in harvesting capacity of the fleet was a major factor in depletion of northern cod in the late 1980s and early 1990s.

Technological equipment such as electronic locating instruments appear to be improving at an increasing rate, possibly exhibiting a positive feedback response as new technologies build on previous advances. This increasing rate of change is of particular concern because in general, technological advances have tended to obscure the effects of fishing activities on fish populations. This can be highly destabilizing because it can permit harvesters to maintain high catch rates even as fish population abundance declines. For instance, in pelagic schooling species such as herring and sardines, the increased efficiency of vessels in finding schools, along with the ability to harvest large portions of them per unit time, means that as fish abundance *decreases*, the proportion of fish caught per unit effort and in total *increases*. This has created a situation where a fish stock can be depleted rapidly, yet the perception based on catch per unit effort might be that the fish stock is still at a healthy abundance. When technological advances occur at an increasing rate, as they appear to be doing, this situation is exacerbated and, if not adjusted for by management regulations, creates risks to both fish populations and to those dependent on those fish.

Another example of masking a decrease in fish abundance may be the adoption of the "turbo trawl" by otter trawlers in the Gulf of St. Lawrence in the late 1980s, which produced a major increase in the amount of fish that could be caught by a vessel per hour. It appears that this change (albeit hard to measure) was not fully taken into account by fishery managers, thus creating an illusion that biomasses of some fish stocks were stable or increasing when in fact they were decreasing. The results were overestimates of biomass levels, which in turn led to quotas and catches that were larger than the stocks could withstand (Harris 1990).

Despite a global trend toward increasing technological sophistication, such change has not uniformly affected all harvesters. For example, this trend has had relatively little impact on subsistence fisheries (such as Native food fisheries), which often continue to use traditional technologies for local food production.

5. Capitalization

In Canada, as in many nations, we have been conditioned to believe that economic growth, through ever increasing investment and ever increasing output, must be positive, even if it is neither sustainable nor in keeping with society's objectives. This belief occurs at an international level, with each nation attempting to increase its output and to outcompete others. It also occurs at a local level; for example, large chains of pharmacies, with access to capital, expand their outlets to such an extent that they force smaller stores out of business within a given community.

It is not surprising then, that this process is also found in capture fisheries, where the tendency is to increase catching and processing capacity through time until it exceeds that needed to harvest at sustainable levels. This may occur with the blessing or even stimulation of government (e.g., excessive subsidies for boat building and fish processing equipment) or due to a lack of appropriate control by government. In either case, fishermen have an economic incentive to invest in more powerful vessels and better gear, so as to catch a greater share of the available limited harvest (see Box 2.6 Fisheries Economics). A common further incentive for this overinvestment and implicit reallocation of catches over time has been the practice of the government management agency of allocating amounts of fish quota among sectors of a fishery and among vessels (e.g., into ITQs) based on their historical share of the catch. Although this may seem a reasonable way to allocate ITQs, it encourages investment in boats and fishing strategies to increase an individual boat's share of the catch prior to the initial allocation of ITQs, even if stocks are declining.

As noted in Chapter 2, Caddy and Gulland (1983) described the tendency for harvesting capacity to increase through time as a "ratchet effect," taking only small and temporary downturns during times when fishing is poor. Those downturns in capacity are usually smaller than the subsequent increases when capital is invested in more fishing-related gear, thus generating the long-term tendency to increase.

Another key point in the context of this report is that when the tendency to increase capitalization through time is taken too far, and it usually is, it has several negative effects: wasted capital, an increased vulnerability of the fishing industry to external economic factors such as price reductions or biological factors that reduce fish abundance, and an increased conservation *risk*. The risk arises if, for example, political pressure to utilize the capacity outweighs concern about reductions in spawning stock biomass. Typically, broad, long-term policy measures are needed to reverse the capitalization process, while in the shorter term, tactical measures are needed in-season by management agencies to control an overcapitalized fleet, for example through effort controls such as limits on types of gear, days fishing, or length of season.

While excess capacity is common in many fisheries, it is not a simple matter to measure its extent. For example, many fisheries are seasonal, so that boats and plants may naturally lie idle at some times of the year, while other fisheries rely on stocks that fluctuate widely. The optimal capacity in either of these situations involves a balance of risks: the "downside" risk of suffering idle capacity in years with low fish abundance (or giving in to demands to use that capacity) versus the "upside" risk of having too little capacity to take advantage of available harvests in good abundance years (Charles 1983). In general, we should expect some idle capacity in bad years and some shortage of capacity in the best years. The exact balance will depend, for example, on our ability to properly manage the fleet and in particular to refrain from using too much capacity in bad years. In any case, it is important to develop proper measures of overcapacity, together with a better understanding of the process of capitalization and its impacts.

Socio-political processes that affect fishery systems

Humans involved in fisheries can also be affected greatly by changes in socio-political factors, which can indirectly influence the sustainability of fishery systems. Profound effects have resulted from changes in the last few decades in how the federal government management agency (most recently the Canada Department of Fisheries and Oceans, DFO) has conceptualized the main policy objectives of management and government's proper role in attaining these objectives. Such changes in the prevailing fisheries management paradigm have led to changes in the distribution of access rights and catch allocations, changes in the definition of subsidies, changes in specialization of fishermen into fewer fisheries, and, on the west coast, changes in the location of fishing licenses and services to less rural settings. Which management paradigm was preeminent at any particular time determined the priority accorded to conservation and risk aversion. Additional changes resulted from the increasing scale of human interactions with the environment. All these changes have also increased the uncertainty about future conditions for fishermen, fishing communities, and the industry as a whole.

1. Changing management paradigms

What have government managers viewed as the chief purpose of management? In any given historical period, there has usually been a reigning paradigm or world view shaping government's vision of the major problems to be addressed, the critical task of fisheries management, and what the underlying purpose of management should be. As new problems and new issues made their way onto the national agenda and as new approaches emerged, competing paradigms gained dominance, or in some cases co-existed in a subdominant position alongside an old or dominant paradigm. This section describes the rise, fall, and interplay of management paradigms from the 1950s through the 1990s. These changes in paradigms have occurred over the decadal scale, or in some cases, longer periods. These paradigms are named below by their chief policy instruments and they fall roughly into the following periods.

(a) Modernization/development: 1950s to early 1970s

Substantial federal and provincial subsidies for vessel construction and technological development, especially focused on capital-intensive vessels owned by processors, were intended to enhance the competitive access of Canada to offshore stocks exploited by distant-water fleets. Early quantitative analyses of fish population dynamics (e.g., Ricker 1954; Beverton and Holt 1957) lent support to development of management philosophies such as maximum sustainable yield, maximum yield per recruit, and maximum economic yield. Quite apart from those analyses, serious depletion was not seen as a potential problem for most fish stocks and governments played the role of developer by targeting specific sectors for receiving subsidies. Processors came to influence fisheries development policies, while fishermen influenced welfare and unemployment policies (MacDonald 1984; Barrett 1984). Although specific practices such as vessel subsidies tapered off in the 1980s and 1990s, the general attitudes surrounding these practices have been slow to change.

(b) Privatization: late 1960s to present

Both before and after Canada gained control of its 200-mile Extended Economic Zone of oceans off its shores in 1977, the modernization paradigm was mitigated by some recognition that overfishing had occurred and that new technologies were not accompanied by enough scientific knowledge to overcome problems in by-catch and multispecies management. However, the main concern of government policymakers and the industry was to reduce fishing costs and to make fishing and processing operations viable in the long-term, without cyclical government bail-outs. Economists first recommended a limited entry vessel licensing system, begun on the west coast in 1969 and extended gradually to the east coast beginning in the mid-1970s. Licenses deemed surplus were bought back or eventually retired by government; most existing fishermen were "grandfathered in," receiving their licenses gratis, but were then able to sell them as a commodity. Licenses sold in this way thus became a limited form of private property. Despite creating fewer vessels, these programs resulted in overall greater fishing power because of increased capitalization (financial investment) by vessel owners. Therefore, government regulation shifted attention to output controls (ITQs: Individual Transferable Quotas). These were first applied fully in 1982 through Enterprise Allocations (EAs) to corporate owners of the east coast offshore trawler fleets and through ITQs for the Bay of Fundy herring purse seine fishery. ITQs and EAs were assumed to create automatic rationalization (reducing overcapacity), although this has not always been the case. In 1996, the FRCC (Atlantic Canada's Fisheries Resource Conservation Council) began discussing the use of more extensive effort controls as a supplement to quota management (including ITQ systems).

In some cases, DFO granted ITQs as part of rationalization. For instance, DFO attempted to streamline east coast processing by creating two east coast super-companies. These companies replaced many smaller fish plants and received fishing ITQs, called "Enterprise Allocations" for their associated vessels (MacDonald 1984; DFO 1976). ITQs are now implemented in offshore and midshore groundfish, crab, shrimp, lobster, herring, and scallop fleets on the east coast, and for halibut, sablefish, and groundfish on the west coast, among others.

But once established, private property rights in fish (ITQs) were considered to be antithetical to detailed regulatory approaches. Guaranteed ownership was assumed to create private incentives to operate efficiently and to harvest sustainably so that government could remove itself from traditional regulatory activity in favor of decentralized management by shareholders in quota-holding companies (McClurg 1997). Because this paradigm envisioned quota owners as paying for government management services, it gained popularity in an era of fiscal restraint when DFO had to recover costs from the industry. DFO's Assistant Deputy Minister, an economist, referred to fishermen who pay for management services as "business clients" (Standing Committee on Fisheries, Feb. 4, 1997). The logical extension of this paradigm would be a substantial reduction in the public policy role played by DFO. Fishermen would be clients paying for all government services and holding the management rights, including the right to do stock assessment and contest the setting of the total allowable catch (as they do in New Zealand). Management decisions would be made almost entirely by private owners of quotas, who hire services and exchange rights through the market, ideally achieving maximum efficiency (at least as narrowly conceived in terms of return to capital investment). The Pacific halibut fishery, widely considered the most successful Canadian ITQ fishery, is discussed in Chapter 6.

(c) Broad-spectrum micro-management: mid-1970s to present

Fisheries minister Romeo LeBlanc decentralized the structure of fisheries management decision-making in the mid-1970s by creating area managers and delegating some decisions to them. LeBlanc thought that the better organized major processors were overly influential in management decisions, so he also made funding available to strengthen fishermen's organizations and encouraged the development of co-management arrangements between DFO and fishermen's organizations (DFO 1976; LeBlanc 1978 as cited in MacDonald 1984; Kearney 1984). DFO decision-making centralized again in the mid-1980s and detailed, area-specific management has been abandoned in many areas, but interest in it jointly by DFO and local NGOs has increased. Initiatives on both coasts have emerged, sometimes with and sometimes without sponsorship and support from DFO. The major thrust of such "co-management" is to craft locally appropriate regulations with considerable input from local fishermen and to use the greater legitimacy of such regulations to pull local associations of fishermen and fishing communities into agreements with government to participate in implementation of regulations. Fishing-dependent communities that become involved may, under appropriate conditions, develop into resource stewards who perform a broad spectrum of management functions (e.g., habitat protection). Thus, co-management also gained popularity partially because it too is seen as a solution to DFO's fiscal crisis (because considerable costs are assumed by local NGOs) and also to crises in the legitimacy of DFO regulations and policies, when widespread public outcry periodically questioned these as non-accountable to regions, communities, or local stock health. Superficially, co-management resembles privatization, in that regulation is decentralized and costs are shared. However, co-management bodies are generally opposed to regulation by the market and to the alienation of access rights from owner–operators in local areas. The 1996–1997 debate over the revised Fisheries Act, Bill C-62, sought to clarify whether the proposed "partnerships" between industry and DFO mean privatization, co-management, or both. A locally initiated and DFO-supported co-management experiment in the Scotia-Fundy region is discussed in Chapter 6.

(d) Conservation/protection: post-1992 on the east coast, post-1994 on the west coast

Although conservation concerns were briefly acted upon after unexpected periods of low fish abundance in the 1970s and 1980s, they had minimal effects on practice. It was not until the 1992 collapse of the east coast cod and the 1994 west coast "missing salmon" situation that managers and the public became sufficiently alarmed and placed the concept of risk averse management rather precariously on the agenda (Fraser River Sockeye Public Review Board 1995). On the west coast, for example, the specter of the Newfoundland cod failure underlay broad public concern for conservation, reflected after 1994 in repeated media and public document statements that Pacific salmon were at risk, and increased calls for a Pacific Conservation Council. This body was conceived as an independent public watchdog acting "for the fish." In May 1996 DFO convened a major workshop on the west coast to discuss with a broad array of interest groups how it might fund a Conservation Council and deal with the rising cost of consulting stakeholders. The workshop reflected changing perceptions of which interest groups were concerned; it included not only an array of environmental groups, but a broad spectrum of other NGOs and community-based groups working on conservation issues. This spectrum of interest groups was also reflected in the province's May 1997 Fisheries Renewal workshop and in a series of university-sponsored public forums during 1995–1997 on fisheries conservation and management issues.

While it has some influence, this conservation paradigm is ambiguously articulated and is not necessarily entrenched in practice. An example of the practice is an expansion of effort into "underexploited" species, notably crab and lumpfish, following the collapse of the northern cod in Atlantic Canada. This shift in effort took place in the absence of sufficient studies or awareness of the true status of these stocks. Landings for crab and lumpfish increased dramatically and there are now serious concerns about their status.

2. The changing concept of subsidies

Related to the above paradigm shifts, there has been a changing concept of what constitutes a subsidy, under what conditions it is desirable, and under what conditions government should recover management costs. This concept is of central importance because subsidies have always been one of the chief instruments of Canada in achieving policy objectives. Stated most simply, government policy has changed radically from large and targeted subsidies to near zero or small targeted subsidies. This shift, which has occurred since about 1989, has fundamentally altered the economics of Canadian fisheries by increasing direct costs to industry.

DFO has increased cost recovery from fishermen by increasing license fees and by requiring that fishermen pay for the cost of many management activities previously performed by DFO. For example, DFO has implemented dockside monitoring of catches in many fisheries. This is done by private companies and the service is paid for by fishermen. Dockside monitoring has also enabled DFO to reduce the number of fishery officers, eliminate their overtime pay, and lay off data entry operators. Cost recovery is still relatively new and raises other public policy issues, e.g., (*a*) are recovered costs distributed equitably?; (*b*) to what extent do they cover public rather than private benefits?; and (*c*) how much are these recovered costs integrated into an accounting of cost-saving measures applied within DFO? These accountability issues are addressed in Chapter 6, but are important to note here as an illustration of the profound shift in DFO's role as funder of all regulatory activity and subsidizer of many changes it considered desirable. This changing role shifts significant costs to industry, creating yet another source of uncertainty for fishermen, along with the associated risks.

3. Specialization

The changing distribution of access rights and allocation described in Chapter 2, as well as the application of new paradigms discussed above, has created a more specialized and capitalized fleet that is more vulnerable to economic downturns, particularly on the west coast. These changes in socio-political policies create increased risks for fish stocks through increased incentives for behaviors by the industry and fishermen that do not favor conservation of fish stocks.

(a) Increased capitalization and debt

Measures adopted to rationalize the fleet, whether by controlling inputs (effort) or by controlling outputs (catch), have resulted in significantly increased debt loads for the majority of remaining fishermen. Limited entry systems capitalized the increased value of the license into the boat, encouraged creative boat construction to increase hold and harvesting capacity, and fueled speculation in licenses and vessels. ITQs capitalized expectations about future benefits into the value of the quota, so that costs of quotas skyrocketed as soon as they were sold by the first generation that had been "grandfathered in" as quota holders. Currently ITQs in Canada sell for two to six times the annual landed value of the catch and approximately half of the quotas in Pacific halibut and sablefish have been transferred (IPHC 1997). The lease, as opposed to outright purchase, of halibut quotas typically costs a leasor two-thirds of the landed value of the catch. In other words, significant profits are being extracted by parties other than the current fishermen and costs to fishermen are high. Increased debts tend to lead fishermen to fish harder and to take greater risks with their own safety at sea and with conservation regulations (Cove 1973; Langdon 1982), leading to greater risks for the fish population. This is particularly problematic for fish stocks where the abundance and allowable catch are highly variable due to natural environmental variation. In such cases, there will inevitably be fishermen whose economic viability will be put at risk, unless the capitalized value of the quotas realistically reflects that variability of fish stocks. This can occur only if concentration of quotas in a few hands is not allowed.

(b) Decreased flexibility and adaptability

As more and more fisheries are brought under limited entry systems, more and more license privileges are removed from vessels that once held them. For example, an "A" license in the B.C. salmon fleet used to include the halibut bycatch and a "C" license covered almost all species except salmon. In a poor season for one species, a fisherman could depend more on other species. However, due to changes in licensing, especially on the west coast, this flexibility has been reduced for the majority of fishermen, who usually now specialize in relatively few fisheries.

The trend to decreased flexibility has varied both spatially and temporally on the west coast. It has particularly affected rural communities on the north coast and the west coast of Vancouver Island, which lost licenses in many cases because they were unaware that new licensing programs were underway and that their privileges would be revoked if they did not land a specific amount of a certain species in particular qualifying years. Entire communities that once depended on particular species have lost their access to them with the introduction of limited entry for salmon in 1969 and for halibut in 1979. For instance, halibut licenses are almost entirely absent from Tsimshian, Nisga'a, Haida, Heiltsuk (Bella Bella), Nuxalk (Bella Coola), and Nuu-chah-nulth First Nations communities.

Single-gear area licensing as introduced in the Pacific salmon fishery in 1997 is a fleet rationalization and concentration instrument rather than a mechanism to tie fishermen to areas. For example, combination gillnet–troll vessels were required to specialize in one gear (e.g., gillnet) and buy out another licensee (e.g., troller) if they wished to continue fishing both gears to obtain a longer or more flexible season in one area. Thus, a fisherman who wanted to fish only one area would have reduced ability to adapt to differing gear opportunities in that area unless he doubled his investment by buying another license ($80 000–$150 000 in 1996–1997). This policy would be beneficial to DFO because it makes more predictable the amount of fishing power that shows up at openings in particular fishing areas, which might permit DFO to meet conservation goals more effectively. However, this policy is disadvantageous for most fishermen in the short run, who can no longer easily "spread the risk" across space and time by simply fishing over a larger space and for a longer time. Trollers who wish to spread risks by fishing in more than one area but who cannot afford to buy a second license have advocated a "gray zone" at the north end of Vancouver Island, which could be used when less abundant stocks are expected in a particular area and year. Such risk-spreading strategies are important for fishermen, because many fish stocks are highly variable in space and time.

In summary, increased debt and loss of flexibility have negative implications for conservation. As well, fleet rationalization measures have serious implications for the economic vulnerability of individual fishermen and communities.

4. Increasing scale and degree of human interactions with the environment

Over centuries, and particularly in recent decades, increases in the human population, amplification of technology, and overall economic activity have increased the impact of humans on natural ecosystems. Furthermore, these changes appear to be occurring at an accelerating rate. As a result, human activities have both global and local effects. On a global level, increases in population and in technology have increased exploitation rates on fish populations to the point where total harvest from wild fish stocks may be nearing a peak (Chapter 2). This intense exploitation threatens the long-term viability and stability of continued harvests because fully exploited or overexploited stocks are more susceptible to even small changes in fishing effort or environmental conditions. On the local level, as fishing capacity and technological capabilities increase, natural refuges for fish are being lost; places where few, if any, fishermen ventured previously are now within reach. For example, this lack of a biological refuge is an issue in the Canadian Atlantic lobster fishery, where many fishermen have expressed concern that collectively they are leaving nowhere for the stocks to escape harvesting pressure. The elimination of refuges and other natural mechanisms that limit fishing mortality on a fish population also reduces the ability of such stocks to withstand unfavorable environmental conditions.

The effects of increasing human activities on fisheries are particularly acute in coastal areas. Indeed, management of fisheries is no longer effective in isolation from other coastal activities. Coastal industries produce increasing levels of pollution, conflicts between commercial fishermen and recreational users of the oceans, conflicts (or even mutual support) between tourism interests and fishing activities, forestry–fishery conflicts, and so on. For instance, coastal aquaculture may have impacts on local fisheries in terms of (*a*) the health of wild stocks (e.g., B.C. salmon) due to the spread of diseases or competition with fish that have escaped from aquaculture facilities, (*b*) the area of the ocean available for fishing activity (e.g., for Nova Scotia lobster), and (*c*) the price paid for wild fish (e.g., in the worldwide market for salmon). Furthermore, salmon hatcheries and other fishery enhancement facilities can influence the relative abundance of different stocks that are harvested simultaneously in commercial fisheries, often leading to overharvesting the less productive wild stocks.

To address the conflicting coastal uses, there has been an increasing move toward "integrated coastal zone management," with the overall goal of balancing among multiple uses of the coastal zone, in order to meet multiple objectives (economic, social, and cultural, as well as ecological). Recent Canadian efforts toward integrated coastal zone management include the Georgia Basin Initiative and the Growth Strategies Amendment Act in British Columbia, Coastal 2000 in Nova Scotia, and the Atlantic Coastal Action Program on the Atlantic coast.

5. Urbanization

Direct effects of increasing human population and urbanization on fish habitat are discussed in a later section. Changes in less direct effects of urbanization also exist. For example, in British Columbia, ownership of a significant proportion of larger fishing boats has shifted away from coastal fishing communities toward major urban centres (see Chapter 2). This affects the sustainability of smaller coastal communities (particularly where a fishery serves as an "engine" for the local economy), because (*a*) such communities have less control over the boats that harvest local resources and (*b*) the employment and secondary income (e.g., money spent on fuel and maintenance) generated from operating the fishing fleet is lost from the community. Furthermore, there could be threats to ecological sustainability if the incentives that local residents and communities have to maintain a sustainable harvest (incentives that can be traced to local dependence on the resource) are less intense among people living a greater distance from those resources. The removal of licenses from hinterland rural communities can also decrease incentives to protect fish habitat from other uses that would be profitable to the local economy.

Effects of biological processes and their interaction with other processes of change

In this section, we provide examples of how changes in the economic, socio-political, physical, and biological processes described above can affect the key biological characteristics of fishery systems: the growth rate, survival rate, and reproductive rate of fish, their spatial distribution and vulnerability to fishing, and the biological structure of aquatic communities. We document mechanisms that explain observed changes in these components.

1. Effects of economic and socio-political processes

Effects of fishing on structure of fish populations

Economic and socio-political processes shape the efficiency, magnitude, or spatial scale of fishing through changes in prices, technological improvements, and incentives. These processes thus affect the biological sustainability of fishery systems because they increase the scale and intensity of many effects of fishing, thereby altering either the harvest component in Fig. 3.1 or the recruitment component indirectly.

For instance, fishing changes the age structure and reproductive output of fish populations by (*a*) increasing the total mortality rate, which decreases the average lifespan of fish, and (*b*) often by removing older individuals from the population (Fig. 3.10). When kept within reasonable limits, the additional mortality caused by fishing is not necessarily harmful because fish populations often have density-dependent mechanisms that result in increased survival, growth, and reproductive rates per individual fish as abundance decreases over some limited range (see Box 2.1). However, the change in the size–age structure of fish populations caused by fishing also affects their total reproductive output. For instance, selective removal of large fish by fishing gear can disproportionately affect the productivity of fish populations because larger, older fish generally produce the most eggs (and probably the most offspring that survive), both per female and per unit biomass of females (Trippel 1995).

This effect of fishing on narrowing the age distribution has serious implications for future biomass yields, particularly for slow-growing and long-lived marine fish species. Various rockfish (*Sebastes*) species on the Pacific coast, for instance, live 80–140 years and do not mature until an age of 10 or more years (Archibald et al. 1983; Leaman and Beamish 1984). Such long-lived fishes by definition have low natural mortality rates and tend to grow in body size relatively slowly (Pauly 1980). Reproductive output by these fishes tends to be spread out over a female's entire lifetime. The effect of these "risk-spreading" or "bet-

Fig. 3.10. Frequency distributions of Pacific ocean perch (*Sebastes alutus*) of different ages from an unexploited stock (top panel) and a heavily exploited stock (bottom panel) off southwest Vancouver Island sampled in 1979 (after Leaman and Beamish 1984).

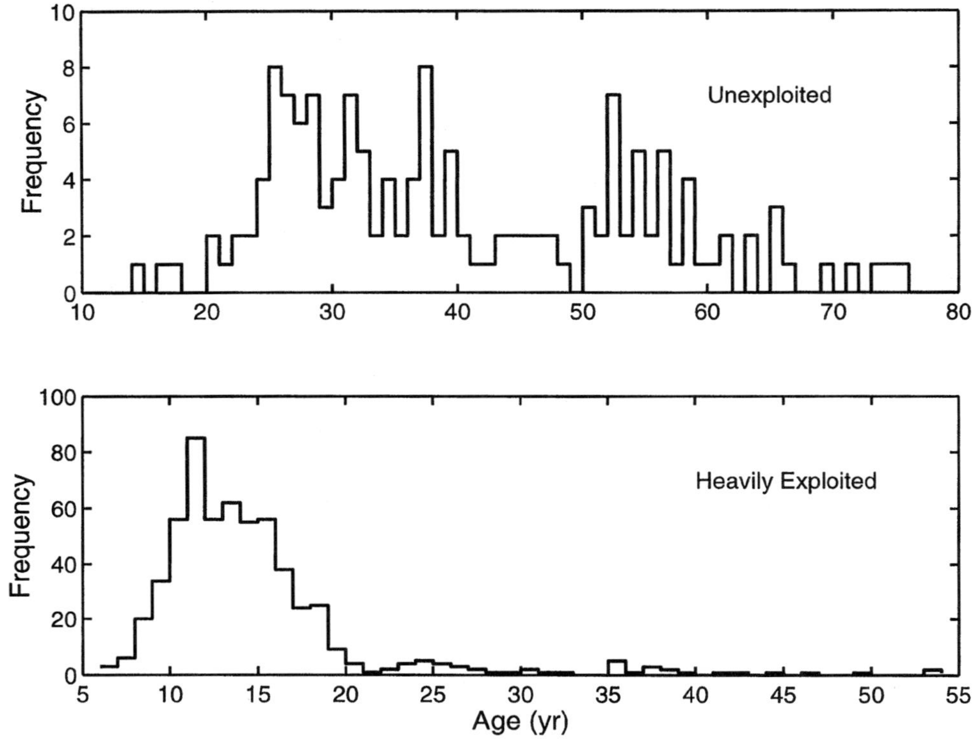

hedging" life history strategies is to maximize the probability that some of a female's offspring will survive in variable environmental conditions. However, fishing undermines the risk-spreading value of long reproductive life by reducing the number of spawning episodes per female and the probability of successful reproduction by any given female. Thus, when fishing is moderately intense and adverse environmental conditions persist, individual females in such populations have a lower probability of reproducing successfully compared to conditions where lifespans are longer. The result is that the fishable biomass of a long-lived fish population that is even moderately harvested is much more likely to decrease in poor environmental conditions than lightly fished stocks (exact definitions of "moderately" and "poor" will depend on the particular species, its density-dependent responses, and its environment).

Furthermore, the higher the harvest rate, the more change there is in a stock's age structure and the worse this effect gets. Also, the longer that intensive fishing occurs, the higher the risks because it results in a longer period during which a stock has to successfully reproduce with an unusually narrow age distribution of spawners. One compensating response that many fish stocks show to reduction in abundance is a lower age at maturity due to increased growth rate of individuals that remain in the population. However, this does not generally completely offset the rapid reduction in abundance caused by harvesting.

In many species of marine fishes, size-selective fishing also frequently increases the mortality rate of younger, smaller fish because, although they may be thrown back into the sea (discarded) due to being too small, their mortality rate is higher than fish that were not initially caught.

Size-selective fishing also has the potential to change the genetic structure of the population over the long term, thereby potentially affecting intrinsic growth rates and age at maturity schedules (e.g., Policansky 1993; Smith et al. 1991).

Effects of fishing on structure of aquatic communities

Not only does fishing change the structure of fish populations, but it also has the potential to alter the structure and functioning of aquatic communities (Gislason 1994), sometimes irreversibly. It is difficult to find unequivocal evidence of this effect, but the weight of current evidence suggests that, by adding more mortality to some species than others, fishing can shift the species composition of marine communities (e.g., Wainwright et al. 1993). This shift occurs either through changing the magnitude of predator–prey interactions or through fishing being an added "competitor" for aquatic resources (Hilden 1997). An example of changing predator interactions occurred in the North Sea where depletion of fish-eating herring and mackerel stocks due to fishing was associated with an increase in the catch (and possibly abundance) of smaller prey species such as sandlance (Hempel 1978, cited in Parsons 1992).

As well, there is substantial evidence for the effects of fishing on community structure on the Georges Bank in the Gulf of Maine (Grosslein et al. 1980; Sissenwine 1986; Sissenwine and Cohen 1991; Fogarty and Murawski 1998). This groundfish community was once dominated by cod and haddock, which are commercially valuable species. However, selective fishing for these two species severely reduced their abundance and the community is now dominated by small elasmobranchs, which are less valuable commercially. It is possible that this change may persist because interactions among juvenile haddock and elasmobranchs may reduce the recovery rate of haddock stocks even if fishing for haddock is eliminated.

The community structure in the Bering Sea ecosystem also changed over the past 30 years. The abundance of herring, whales, fur seals, sea lions, and most crab species decreased, while abundance (not just catch) of pollock has greatly increased (NRC 1996). The United States National Research Council Committee on the Bering Sea Ecosystem (NRC 1996) concluded that the observed increase in abundance of pollock resulted from a regime shift in the physical environment, combined with intense harvesting of various species that either competed with, or fed on, juvenile pollock, such as whales, yellowfin sole, Pacific Ocean perch, and sablefish.

Effects on habitat

Economic and socio-political processes can also affect fish through changes in habitat. This is a particular concern in B.C. for Pacific salmon, where individual streams support distinct stocks and where activities such as urbanization and logging affect streams through burial or siltation (e.g., Magnuson et al. 1995). Increased economic development and urbanization of coastal areas may reduce survival rates of certain fishes by compromising spawning, feeding, or rearing habitat. Increased concentrations of pollutants in coastal waters from urban development may reduce survival rates of fish (e.g., effect of wood preservatives on salmon smolts — Kruzynski and Birtwell 1994; also see Griswold 1997).

Fishing can also have an indirect effect on the structure of the biological community by changing the habitat. This was demonstrated by a 5-year, large-scale experiment off the coast of northwestern Australia, which showed that intensive trawl fishing had shifted the groundfish species composition to less commercially desirable species by removing the large benthos (sponges and other large invertebrates) that made up the preferred habitat of the more desirable fish species (Sainsbury et al. 1997). Experiments in other habitats are underway but as yet have not produced such clear conclusions (e.g., Prena et al. 1996; Morgan et al. 1997).

2. Effects of physical environmental processes

Physical environmental processes also affect fish populations. Even in the absence of commercial harvesting, the abundance of some fish populations can be quite variable over either short or long periods (Beverton 1993). For example, the abundance of Pacific sardine and anchovy off the coast of California has been estimated over the last 1700 years using counts of fish scales in sediment cores taken from the ocean floor (Baumgartner et al. 1992; Fig. 3.11). While these data only provide an index of abundance over a relatively small spatial scale, they suggest that abundance of both species has been highly variable. Most of these data predate intensive commercial fisheries, hence they suggest that physical or biological processes alone can potentially cause significant fluctuations in abundance of fishes.

The next few sections provide examples of how physical environmental processes affect key biological processes in fishery systems at four time scales.

Long-term changes in physical processes

A long-term change in climate and related changes in the speed and direction of winds, currents, and vertical mixing of the ocean can potentially affect growth and survival rates of fish (Ware and Thomson 1991; Brodeur and Ware 1992). For instance, wind-induced upwelling affects abundance of zooplankton, which are the primary food source for many fish species in coastal upwelling zones (Bakun 1990). However, the effect of increased plankton availability on fishable biomass of commercial fish species is not known definitively because of the difficulty in sorting out confounding effects in historical data and identifying relationships between changes in productivity and biomass at several successive levels in the food chain. Thus, and most importantly from the perspective of this report, while past changes in ocean climate on a decadal scale and through NAO and ENSO events have resulted in changes in distribution, survival rates, and growth rates of fish, there is at present no clear basis to forecast what *future* climate change will do to these variables (IPCC 1996). Because of the differences among regions of the oceans in responding to climate change, the IPCC (1996) report concludes that there will *not* likely be major changes in total *global* biomass of fish stocks. However, there may be some effect on *regional* spatial distribution of fish populations, which may in turn affect the relative abundance and species composition of fish communities in some areas. In this context, Frank et al. (1990) speculated that long-term climatic change would lead to a general warming and freshening of coastal Atlantic waters and predicted northward shifts in spatial distribution of some groundfish stocks, changes in migratory behavior of some pelagic fish species, and a trend towards replacement of groundfish communities with pelagic fish communities. These changes would have significant implications for those human communities that rely on local groundfish stocks. However, the uncertainty in the prediction of such future changes is large.

Finally, long-term changes in temperature and currents could also affect commercially valuable species. For instance, because of their different thermal tolerances, the inshore distribution and subsequent inshore catches of Atlantic cod and Atlantic mackerel can be influenced by the shoreward advection of warmer surface waters (Rose and Leggett 1988; Castonguay et al. 1991). Although the distribution and short-term movements of cod in the

Fig. 3.11. Estimated biomass of Pacific sardine (top) and northern anchovy (bottom) off the coast of California. Estimates were derived by regression from scale counts in sediment cores and go back about 1700 years from the present (after Baumgartner et al. 1992).

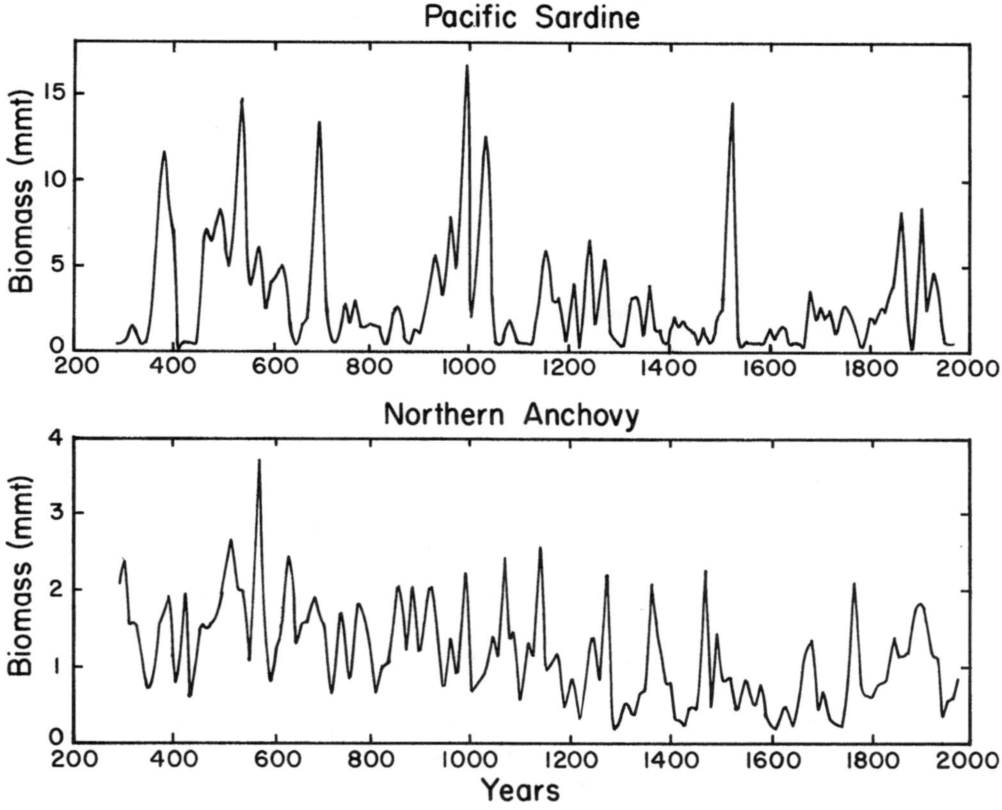

North Atlantic are influenced by temperature, we do not yet know what the response will be on a very long time scale, although there has been some speculation (Frank et al. 1988).

Regime shifts or decadal-scale changes in physical processes

Regime shifts in physical oceanographic conditions in the North Pacific likely affect fish populations through the distribution and abundance of their food supply (Beamish and Boullion 1993; Francis and Hare 1994; Brodeur and Ware 1992, 1995). Chelton (1984) and Pearcy (1992) hypothesized one mechanism that could cause such changes: shifts in oceanographic currents could divert zooplankton-rich waters away from southern regions of the northeast Pacific (southern B.C., Washington, and Oregon) and towards northern regions (northern B.C. and Alaska). Brodeur and Ware (1992) hypothesized another mechanism, in which altered ocean currents brought nutrient-rich water to the surface in the Gulf of Alaska gyre, leading to increased productivity. These hypotheses are consistent with the observed large increase in zooplankton biomass in the Gulf of Alaska between the late 1950s and late 1980s (Brodeur and Ware 1992). It is unknown whether this change is associated with climate change or natural variability of the North Pacific.

Unfortunately, there is no *direct* evidence relating short-term changes in survival rates of fish to changes in food supply because extensive data on marine food supply over space and time are scarce. However, there is indirect evidence. There have been large increases in recruitment and catch of groundfish (Hollowed and Wooster 1995) and catch of Pacific salmon (Beamish and Bouillon 1993; Francis and Hare 1994) since the most recent environmental shift in the northeastern Pacific in the mid-1970s (Fig. 3.12). This suggests indirectly that survival rates may have changed over the same period as the observed increased in zooplankton abundance noted by Brodeur and Ware (1992), or that spawner abundance has increased, or both. More direct measures of fish survival rates also show this pattern — an index of survival rate of Alaskan sockeye salmon stocks, when corrected for the influence of spawner abundance, showed large increase after the mid-1970s (Adkison et al. 1996; Peterman et al. 1998). Only a few sockeye stocks in B.C. showed any significant change and Washington and Oregon salmon stocks have experienced the opposite of Alaskan sockeye — reduced survival — leading to the closures to harvesting in the mid-1990s. While there may also have been some changes in survival rates in freshwater habitats, these geographic patterns in survival rates are consistent with the various hypotheses concerning the effects of climatic processes on the distribution of zooplankton-rich waters. The post-mid-1970s pattern of unusually high production by Alaskan sockeye appears to have persisted as long as, if not longer than, any previous such period (Francis and Hare 1994). If previous mechanisms still apply, there is a high probability that there may be a shift back to the less productive regime in the near future. For those few British Columbia stocks (e.g., Skeena River sockeye) that appear to have become more productive since the mid-1970s, this change may reduce their survival rate in the ocean and hence the harvest that can be expected from them. For Alaskan sockeye stocks, there is likely to be a larger decrease in productivity and harvests, perhaps offset somewhat in economic terms by increases in price per kilogram of salmon.

Helle and Hoffman (1995) showed reductions in age-specific body size of Alaskan and Washington stocks of chum salmon from 1975 to 1992 and concluded that changes in oceanographic conditions in the Pacific during this period may have been partially responsible for the observed decline in body size. However, in both regions, evidence for environmental effects alone was not conclusive because the large increase in wild Alaskan salmon abundance and in salmon released (over 2 billion per year) from North American, Russian, and Japanese enhancement facilities during this same period could have reduced growth rates through competition for food (Rogers and Ruggerone 1993). Clearly there is a potential for interaction between the environmental processes that drive productivity of food for important harvested species and the change in abundance of the fish, whether due to artificial enhancement or natural processes.

Another example of decadal-scale changes in environment and fish comes from the range and abundance of West Greenland cod, which show a remarkable correspondence with local temperature (Brander 1995). Large increases occurred in the 1920s and 1950s (Fig. 3.13). Weight at a given age also changed in synchrony with temperature, as is observed for cod elsewhere in the North Atlantic. The correlation between cod catches and temperature holds up until the 1970s. There is an established connection between Iceland cod and Greenland (Schopka 1994), which is part of the explanation for the observed increases of cod in West Greenland; however, overfishing probably also played a role in the declines. Up until the collapse of the northern cod, there appeared to be a strong relationship between salinity on the Newfoundland shelf and recruitment. No clear model exists for the effect of temperature on abundance of northern cod, although there may be an effect on

Fig. 3.12. History of catches (millions of fish, dashed line), intervention model fit (thin solid line), and estimated interventions (thick solid line) for western Alaska sockeye salmon catches (after Francis and Hare 1994).

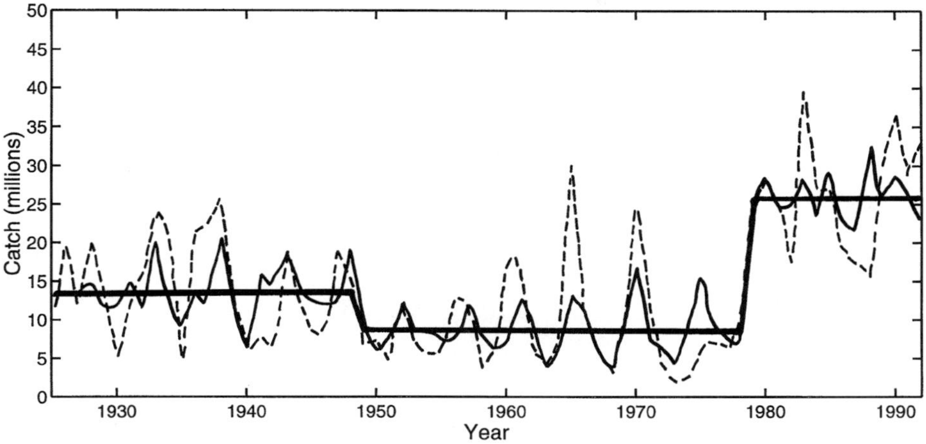

distribution (deYoung and Rose 1993, but see Hutchings 1996 for an opposing view).

Changes in oceanographic conditions can also affect the structure of fish communities by changing the competitive balance and relative abundance of different species. This process may explain the observed variability in relative abundance of California sardine and northern anchovy (Fig. 3.11). Pacific sardine changed abundance dramatically every 60 years or so, whereas the northern anchovy changed roughly every 100 years.

Interannual physical events

Survival rate of fish is also affected by interannual (between-year) changes in the ocean environment. In many long-lived marine groundfish species, recruitment can vary by as much as a factor of 10–1000 between successive years, even though spawning stock biomass changes much less than that over a one-year period. Therefore, such large variations in recruitment must be due mostly to large between-year variations in egg production and (or) in the survival rate of offspring. Egg production per unit biomass does not vary nearly as much as recruitment, leading to the conclusion that variation in survival rate of eggs is mainly responsible for the observed variability in recruitment. This variation is particularly important for species where recruits constitute a significant portion of the spawning and fishable biomass (e.g., short-lived pelagic schooling species such as herring and northern anchovy and some Pacific salmon species such as sockeye, pink, and chum).

Important evidence of the effect on fish of physical oceanographic changes comes from the extreme events associated with the El Nino-Southern Oscillation (ENSO) and the North Atlantic Oscillation (NAO). There are numerous reports of anomalously high or low recruitment or stock biomass during ENSO years for species ranging from herring to groundfish and salmon on the Pacific coast (e.g., Wooster and Fluharty 1985). ENSO events result in northward movements of warm water along the west coast of North America. These warmer waters are relatively low in nutrients and zooplankton abundance (Wooster and Fluharty 1985; Mullin 1995), but are rich in species such as Pacific hake and Pacific mackerel, which are predators of commercially important fish stocks off the coast of B.C. This suggests that reductions in survival rates could be caused by decreased food abundance during ENSO

Fig. 3.13. Cod catches and ocean temperature at West Greenland. Temperature data are running 5-year means of surface values up to 1972 (from Smed 1980) and running 5-year means from Fylla Bank (Buch et al. 1994). Local catches are shown for the period from 1912 to 1937 because they reacted more to local changes in abundance. The local catch is scaled to the international catch from ICNAF Division 1 in West Greenland by a factor of 20.

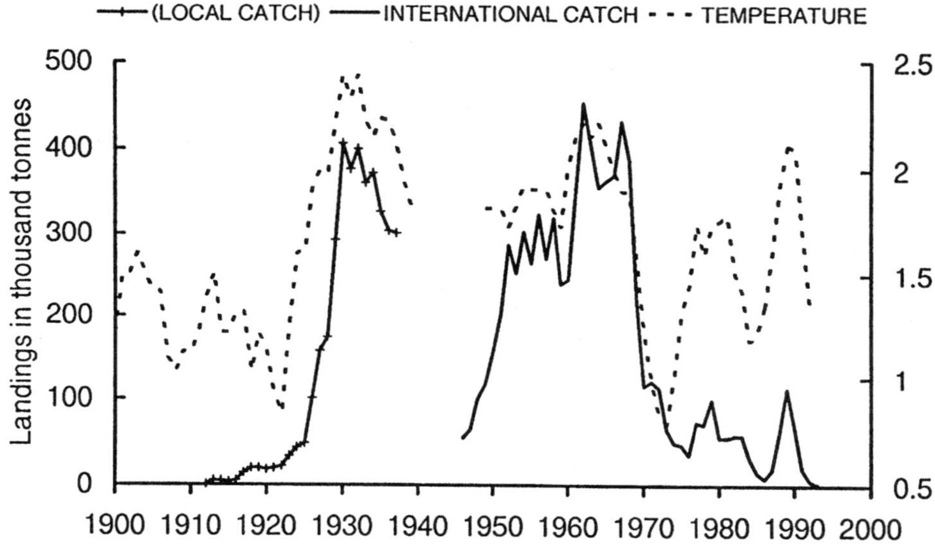

years and (or) by increased predation. For example, Ware (1991) found that recruitment of Pacific herring is strongly reduced by predation by Pacific hake. These observations provide some evidence for ENSO-induced reductions in survival rate and recruitment, but it is often difficult to distinguish between ENSO-induced fluctuations and those caused by other normal aspects of fish population dynamics (reviewed by Mysak 1986).

In addition, interannual physical events can affect fish growth rate. For instance, in the Atlantic, connections have been made between variability in the North Atlantic Oscillation and abundance of zooplankton (Taylor 1995) and between NAO and growth and distribution of cod (Brander 1994; deYoung and Rose 1993). In the Pacific, ENSO events have reduced growth rates of several species of fish (VenTresca et al. 1995; Ware 1991) and there is generally a decrease in body size of fish at a given age if fish encounter ENSO events early in their marine life history. For instance, adult Fraser River, B.C. pink salmon, which only mature as 2-year-olds, were unusually small in the year-class that entered the ocean during the spring of 1958, even after accounting for density-dependent effects on growth (Peterman 1987). This is possibly because coastal waters in southern B.C. in 1958 were low in zooplankton due to the northward intrusion of warm, zooplankton-poor water from the 1957–1958 ENSO event. Many other papers report smaller-than-average body size of fish at a given age as a result of the 1982–1983 ENSO event (e.g., IPSFC 1984; Pearcy 1992), but those observed reductions in body size at age are not necessarily attributable to the ENSO because they could be at least partly due to increased fish abundance.

Productivity of regional sardine stocks along the Spanish coast appears to be related to wind forcing and there has been some speculation about the possible changes that could occur at longer time scales (Bakun 1997). Upwelling along the coast of Spain, driven by winds measured near Cape Finisterre, show an increasing trend from 1950 to 1980 (Dickson et al. 1988). Surprisingly, increased upwelling in spring, which should lead to increased plankton production during the spawning season, is negatively associated with sardine catches in that region. In other upwelling areas of the world, the correlation is generally positive, showing that the same physical process can produce very different biological responses depending upon the specific linkage between the life history of the organism and the oceanic environment. This difference may be explained by the observation that too strong upwelling may carry sardine eggs and larvae offshore away from high concentrations of food, whereas too weak upwelling may produce little food for sardines (Cury and Roy 1989).

Finally, spatial patterns of migration of fish can be affected by oceanographic conditions. Figure 3.14 shows that in the latest time period compared to pre-1970s, a larger portion of adult sockeye salmon stocks migrated on their way to the Fraser River in B.C. around the northern end of Vancouver Island through Johnstone Strait. Mysak (1986) reviewed evidence indicating that the proportion of fish migrating via the northern route was associated with water temperature (reflecting ocean current patterns). The migration route is particularly significant because when these fish come through Johnstone Strait, they are harvested mainly by commercial vessels in Canada, whereas if they migrate around the southern end of Vancouver Island, a larger portion of these stocks is caught by American fishermen.

Vulnerability (catchability) of fish can also be affected by ocean temperature and currents. For instance, in 1972, the Peruvian anchoveta stock became concentrated near the coast due to an ENSO event (Pauly and Tsukayama 1987). High concentrations of anchoveta resulted in unusually high catch per unit effort (CPUE) and managers prolonged fishing openings despite warnings that the stock was extremely vulnerable to overharvesting. While such increased availability appeared to exist, it did not reflect actual stock abundance. The stock abundance declined precipitously and stayed relatively low for many years before beginning a strong recovery.

This collapse of the Peruvian anchoveta has been debated extensively. Some scientists argue that the stock would have collapsed anyway due to changing environmental conditions, even without fishing. Others argue that fishing greatly increased the probability of a collapse and made it collapse sooner than it would have without fishing. Still other scientists suggest that the stock collapsed due to the combination of unfavorable environmental conditions and the effects of fishing on age structure (and the resulting reduced reproductive potential). Regardless of the outcome of this debate, it is clear from our review that both environmental factors and effects of fishing must be taken into account when evaluating strategies to maintain productive fisheries.

ENSO and NAO events constitute only one extreme type of interannual variation and they only occur every few years. Correlations between observed environmental changes and components of fishable biomass also provide information, but it is limited. Such correlational, as opposed to experimental, studies can *suggest* possible mechanisms, but provide little conclusive evidence that these mechanisms actually occur (Walters and Collie 1988).

In spite of these limitations, there is evidence for some relationships between interannual environmental variations and biological processes. One example is the effect of upwelling on survival of juvenile coho salmon off Oregon (Pearcy 1992). These juveniles tend to congregate in nearshore "upwelling zones" during their first months in the ocean and historically, their between-year variation in survival rate was positively correlated with the strength of upwelling. However, this relationship broke down recently for unknown reasons.

Fig. 3.14. Percentage of adult sockeye salmon that migrated through the northern (Johnstone Strait) migration route on their way toward the Fraser River, B.C. and other southern spawning grounds during 1953–1994. The figure from Mysak (1986) for years 1953–1984 was updated with unpublished data from Skip McKinnell (DFO, Pacific Biological Station, Nanaimo, B.C., personal communication, 1997).

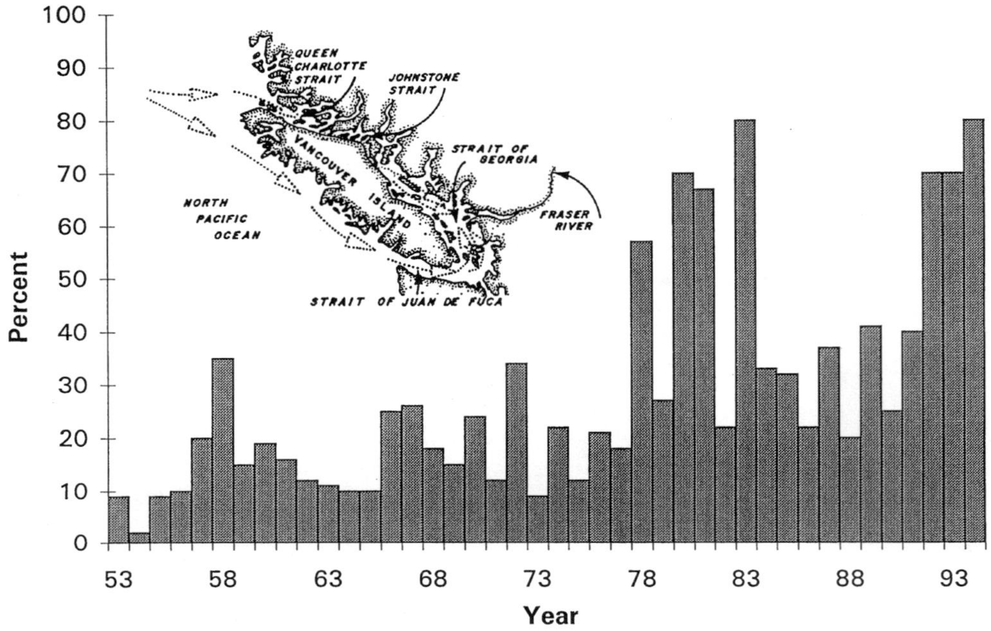

Comparisons of the interannual variability in recruitment of several nearby or overlapping stocks and species also strongly suggest that oceanographic processes affect fish. Results from several studies (e.g., Thompson and Page 1989; Hollowed and Wooster 1992; Peterman et al. 1998*a*) show positive correlation in recruitment (or its components) among several stocks. For instance, Hollowed and Wooster (1992) found patterns of covariation in strong and weak annual recruitments among many groundfish species in the North Pacific (Fig. 3.15). Peterman et al. (1998*a*) also showed that survival rates of sockeye salmon to adult age tend to be positively correlated *within* a group of sixteen Fraser River, British Columbia sockeye salmon stocks and *within* nine Alaskan sockeye stocks from Bristol Bay, but not *between* sockeye stocks from those two regions, even though they overlap for most of their marine life in the Gulf of Alaska. In contrast, there are strong positive correlations in body size of 4-year-old adult sockeye salmon *both within and between* those regions (Pyper et al. 1999). These correlations suggest that interannual variation in growth rate is affected by environmental conditions common to the covarying stocks, but that variation in survival rate mainly arises in regions of the ocean where sockeye salmon from Alaska and B.C. are uniquely distributed, i.e., near their coasts. Myers et al. (1995) similarly concluded from the spatial scale of correlations in recruitment for cod and haddock that regional-scale environmental conditions affected recruitment, rather than conditions common across the entire North Atlantic.

While such regional patterns in survival rate or annual recruitment may be related to oceanographic processes, they may also be partly due to density-dependent factors encoun-

Fig. 3.15. Summary of the percentage of northeast Pacific groundfish stocks that had extreme year-classes. Years are placed above or below the bars when 50% or more of the stocks were extreme (after Hollowed and Wooster 1992).

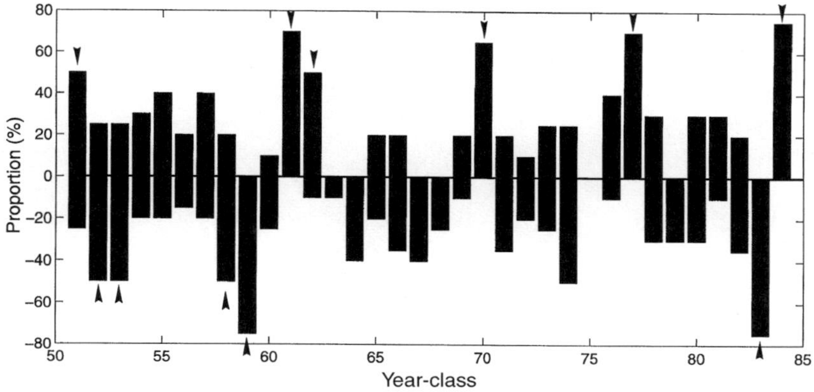

tered simultaneously by different fish stocks. For example, several authors have documented density-dependent growth rates in Pacific salmon (Peterman 1984; Rogers and Ruggerone 1993; McKinnell 1995), which led to reduced body size at a given age as abundance of competitors increased. These density-dependent processes can also act in concert with environmental effects to produce observed changes in body size. For example, ocean temperatures affect the oceanic distribution of Pacific salmon (Welch et al. 1995), which may produce changes in the density of fish in the open ocean and therefore may affect body sizes due to density-dependent growth rates.

Short-term changes in physical processes

Environmental factors that change over short periods can also potentially produce large interannual variation in abundance of recruits of some fish species (Bradford and Cabana 1996; Rothschild 1986). The effects of short-term physical processes on fish depend on the species' life history and on the spatial and temporal dynamics of the ocean. Atlantic cod eggs, for example, after release at the bottom, rise to the surface and within a few days develop into larvae. These larvae are then subject to the currents and temperatures at the surface. Scientists believe that cod have adapted to this environment and spawn in locations that maximize chances of success for the larvae. But this is only true on average. For any given year, month, or day, a storm could occur and generate currents that would carry the larvae far from their preferred location (Sinclair 1988). Also, the smallest scale mixing process, turbulent movement, is important in bringing the larvae into contact with their prey (Rothschild and Osborn 1988). Both of these physical effects are influenced by wind. Thus, while average conditions may be beneficial for larvae, a series of events such as storms, or perhaps even a single storm, can lead to the loss of a group of larvae (Lasker 1976; Peterman and Bradford 1987). The collection of such events in a single year could then be responsible for recruitment failure.

Single pollution events such as oil or other chemical spills could also be considered under short-term physical changes, although there may be long-term biological effects as well.

Common characteristics of processes of change

Previous sections have emphasized the multi-faceted economic, socio-political, physical, and biological changes that affect fishery systems. These changes produce serious challenges to creating sustainable fisheries because the resulting variability and uncertainty add to the complexity of the total fish–human system. Thus, attempts by management agencies to develop appropriate regulations, as well as by industry and communities to develop appropriate investment plans, *must* take this variability and uncertainty into account (see Chapters 5 and 6 for how to do so).

One further useful step, then, is to discuss some simple, but important, common characteristics of these processes of change: spatial heterogeneity, temporal variability, difficulty in detecting the changes, and their potential irreversibility (at least in the short term).

1. Spatial heterogeneity

We have seen that factors that affect fisheries have various geographical extents, have various geographical extents. For instance, on the smallest scale, increased frequency of high local winds in one year can reduce feeding success and survival rate of a year-class of juvenile fish through dissipation of concentrations of their food in surface waters. In some cases, this may reduce the abundance and future catch of fish. On a larger, regional scale, shifts in management regulations can alter the movement of vessels among fishing areas or species. Regional weather and ocean conditions can affect ocean currents. For instance, the regional atmospheric and oceanographic changes that occurred in the North Pacific in the mid-1970s led to an apparent increase in the productivity of Alaskan sockeye salmon stocks, expanding fishing opportunities for harvesters in that area, but decreasing them off the coasts of Washington and Oregon. On the largest scale (global), shifts in prices and technology, climatic change, and globalization of markets create further uncertainty for users and management agencies. Thus, recognizing the spatial scale at which a particular process of change occurs improves our understanding of its effects and may help suggest ways to either influence it or compensate for its effects.

2. Temporal variability

We have also seen that processes affecting sustainability of fisheries can be quite variable on several time scales. For instance, it appears that fisheries existing in environments subject to wide variations in productivity (on any time scale) may have to be severely restricted, or even closed occasionally. Such limitations may be necessary for lengthy periods in the case of environments that undergo "regime shifts" in productivity. While proper timing of such closures would help ensure the persistence of the fish stocks in question, closures *could* be disastrous for fishermen and coastal communities, at least in the *short* term. However, the alternative, of *not* closing fisheries when productivity becomes extremely poor, will *certainly* be disastrous over the *long* term because without fish stocks, there will be no economic or social benefits from fishing. The strong possibility of such lengthy closures means that adequate risk-spreading economic and social structures must be developed to permit the fishing industry and fishing-dependent communities to be resilient to changes on the space and time scales of the fish. In other words, ways need to be found to enable these people to spread the risks across time, space, resources, and other dimensions. Such risk-spreading strategies are commonly used by managers of financial investment portfolios and animal populations to persist in highly variable and unpredictable systems, quite analogous to aquatic ecosystems. These ideas are elaborated in later chapters.

A key related observation is that some problems in fisheries are created by differences in the human and biological time scales of variation. For instance, consider the interaction between the duration of major regime shifts in the environment that affect productivity and the duration of capital investments (and their payback) in new vessels, equipment, or processing capacity. As described previously, unusually productive environmental regimes typically persist for about 10–15 years and then revert to less productive regimes. However, this time scale of environmental variation may tend to make the process of paying back loans (which tend to be even longer term) out of phase with the productivity of the resource, with serious consequences for fish stocks, fishermen, and their communities. For instance, by the time that improved productivity of some stock due to a regime shift has been detected and acted upon by managers and industry, a highly capitalized fleet may face within a short period a stock with greatly reduced natural productivity, quite apart from the effects of fishing. This asynchrony of time-varying physical, biological, social, and economic processes may be at the heart of many problems that face Canadian and other fisheries. Another related issue is that in some cases, the fishing industry is seeking multiyear fishing plans, with some minimum level of harvest. But clearly, in the context of the changes and variations discussed in this report, this is not a reasonable expectation if sustainable stocks are desired. Flexibility has to exist to prevent overharvest when environmental conditions become unfavorable.

But in natural systems, processes do not exist on all possible spatial and temporal scales of change; they are generally correlated — short time-scale processes usually occur at small spatial scales, whereas long-term processes usually occur at large spatial scales (Steele 1985). For example, small-scale weather events can change in a few days while changes in climate usually take decades or centuries. However, there are very important exceptions to this general rule. Some economic processes that are global, such as fluctuations in prices or exchange rates, can also change in a few days or months. Similarly, large-scale oceanic mixing processes can potentially change relatively quickly (in less than a decade, Manabe and Stouffer 1995). These exceptions that lead to rapid changes have significant implications for fisheries management agencies, the industry, and fishing-dependent communities. Therefore, managers, industry, and coastal communities should anticipate and plan for possible rapid changes in global processes. For instance, it is inappropriate for the fishing industry to make long-term investment decisions based on the average amount of fish caught over the last five years because the productivity of fish can change rapidly and by large amounts.

3. Irreversibility

Changes in factors that affect fisheries can result in irreversible changes in the biology of fish. For example, extended periods of poor oceanographic conditions or excessive harvests may reduce the abundance of a fish stock to a point where it is unable to recover, even if oceanographic conditions improve or if fishing is eliminated. Similarly, changes in the species composition of fish communities caused by climate change or selective harvesting can also be irreversible. Therefore, managers and users of fisheries should not assume that the abundance and harvest of a depleted stock of fish will return to its previous state even if the cause of the decline is eliminated. Management and investment plans should consider the possibility that such changes in fish stocks may be permanent. Even if these changes are only for several decades (as shown by the sardine scale data in Fig. 3.11), this downturn may be too long for industry to withstand.

Many other processes discussed in this chapter are irreversible, at least in the short term, and participants in fisheries should consider this possibility. This is particularly true for global processes such as the integration of world markets or liberalization of trade. However, local changes in administrative arrangements or organizational structures appear to be more reversible (although their past effects on resources may not be). For example, responsibility for managing inshore fisheries in the South Pacific island of Fiji has recently reverted from a centralized management agency to local management authorities. Other processes, including most physical and biological processes and changes in consumer demand, vary between years, frequently reversing direction but not in a controllable or predictable manner.

4. Hard to detect

Changes in factors that affect fisheries are often difficult to identify because of natural variability and our inability to measure fishery systems perfectly. Detectability of changes is also made more difficult when the spatial or temporal scales of the change do not correspond to the spatial and temporal scales of sampling programs. For instance, long-term trends in average global temperatures have been obscured by large between-year variations in annual temperatures (IPCC 1996). This between-year variation also obscures relationships between temperature and biological processes such as growth rates, precluding a clear understanding of the potential consequences of long-term trends in temperatures. Economic and social changes, such as changes in technology, markets, or the socio-economic structures of coastal communities are also variable over space and time, making them difficult to detect until after large changes have occurred.

This difficulty in detection of changes creates problems for managers and users of fish. When changes in technology leading to increased harvesting capacity are not noticed or acted on by management agencies, catches in the short term may be higher than anticipated. Conversely, a change in ocean conditions may increase productivity of fish. If this increase in productivity is not detected, catches of fish may be smaller than they could have been if the change in productivity was noticed and incorporated into management plans. In either case, the inability to detect many processes of change can result in incorrect management or investment decisions and can prevent the sustainable harvest of fish stocks.

Conclusions

This chapter has demonstrated that a main characteristic of Canadian marine fishery systems is that they are dynamic, not static, because many biological, physical, economic, and socio-political processes undergo changes. Some of these changes occur rapidly and lead to large year-to-year variability, whereas others occur slowly or persist for long periods in new conditions. Some processes of variability and change interact. For example, the reduction in mean size of fish populations resulting from fishing tends to make stocks more vulnerable to unfavorable environmental conditions. These pervasive sources of variability and change in fishery systems create uncertainties not only about future conditions, but also about interpretations of past events. The next chapter discusses the implications of these uncertainties.

Chapter 4. Uncertainty and Risk

Introduction

Our knowledge of the physical, biological, economic, and socio-political components of fisheries is limited, partly due to their complexity and partly to the change and variability over time inherent in such systems (Chapter 3). The resulting uncertainty is pervasive, has always been present, and has direct implications for management of living marine resources and for the response to change by human institutions and coastal communities. Uncertainty in any component of fisheries systems can potentially affect other components through the complex linkages described in Chapter 3 (Fig. 3.2). For example, uncertainty about the potential effects of climate change on winds, currents, and water temperature or survival rates of fish stocks necessarily results in uncertainty in forecasted trends in abundance of fish populations and hence in appropriate harvest levels. These uncertainties also affect investment or other economic decisions. Uncertainty in how harvesters will respond to management measures adopted in response to a changing environment can lead to ineffective regulations that may fail to achieve the desired objectives, resulting in adverse impacts on fish resources and the human communities dependent on them.

Chapter 3 discussed how various sources of change and variability in economic, socio-political, biological, and physical processes interact to create uncertainties. These changes can be thought of as causing natural, inherent variability (stochasticity) in fishery systems. In this chapter, we first discuss two other fundamental sources of uncertainty: structural uncertainty and measurement error. Structural uncertainty refers to the lack of knowledge about how processes in fishery systems work; for instance, this uncertainty reduces the ability to forecast responses of fishermen to regulations. Measurement error includes limits to the empirical information available for fishery systems, as well as the precision and accuracy with which economic, biological, social, and physical components are measured. Here we use "measurement" in the most general sense of gathering data and estimating quantities. In fishery systems, this is done in most cases through sampling programs that collect data at specified times and places. For logistical and economic reasons, the number of such samples is limited but the statistical design of sampling programs attempts to ensure that the samples reflect the full system. "Precision" refers to the repeatability of successive estimates of a specific measurement quantity; the more similar each successive value is, the more precise the estimate is (note that a "quantity" may also change for other reasons). A fundamentally different aspect of measurement, "accuracy," refers to how close an estimate is to the "true" value. A method of estimating the annual mortality rate of fish due to fishing, for example, may be relatively precise if many estimates by the same method on the same population come close to the same value, but the mean estimated mortality rate may be biased low (underestimated) if, for example, sampling did not take into account the fish discarded at sea.

Variability due to intrinsic stochasticity is unavoidable, although we may be able to uncover some causes of this type of uncertainty. Uncertainty due to inaccurate measurements or sampling variability can, in principle, be reduced to some extent by improvements in technology, statistical design, or increases in sampling effort. Similarly, we can expect advancement in our understanding of the structure and dynamics of the system with increased study and accrual of information, leading to a reduction in structural uncertainty.

Regardless of their sources, these uncertainties translate directly into risks in resource management (see Box 4.1). After discussing this topic, we then go on to describe possible approaches to risk assessment (estimating their probability and magnitude) and risk management (how to make decisions in the presence of uncertainties and risk). Recognizing and coping with uncertainty requires a different way of viewing fishery systems from the traditional approach. In particular, our perspective must change from one based on point predictions of the effects of environmental variability, management actions, or institutional change to forecasts described in terms of probabilities of different possible outcomes. This requires a change in the way in which scientific advice is formulated and conveyed. We will see that scientific advice in support of fishery management is increasingly being cast in a probabilistic framework and that proper use and interpretation of information on uncertainties and risks requires participatory decision making by everyone involved in fisheries.

Box 4.1. What is Risk?

People generally define risk intuitively as the chance (or probability) of "some bad event happening." However, risk assessors frequently define it by taking into account both the probabilities of different events occurring (good or bad) and the associated magnitudes of the effects. Such measures of risk are important because, for instance, they indicate for decision makers that even though there might be only a 10% chance of causing a fish population to go extinct with a given management regulation, the magnitude of that loss might make that regulation unacceptable. Thus, estimates of risk must take into account both components — the probabilities and consequences of events.

Sources of uncertainty

Uncertainty arises at several points in the dynamics of fishery systems (ellipses in Fig. 4.1, which we discuss below). In our conception, a "fishery system" includes physical, biological, economic, and social components, all of which change through time (Chapter 3). However, our observations do not perfectly reflect the "natural system" because data are collected by sampling a subset of variables and data are subject to natural variability and measurement error (Fig. 4.1). We also commonly collect data on numbers of fishing licenses, sizes of boats, and tonnage, as well as numbers of fish of given ages in the catch. Methods of analysis attempt to synthesize such data into estimates of variables and parameters that describe some measure (e.g., abundance of spawning stock or number of fishing vessels of a certain size) or processes of change (survival and growth rate of fish). However, errors in estimating such quantities are inherent in any method of estimation. Furthermore, our models (whether conceptual or quantitative) are always uncertain about the structure of the system (e.g., the mechanisms causing prices to change or the fish stock-recruitment relationship). This structural uncertainty can thus lead indirectly to incorrect choices by decision makers in industry and management agencies.

Analyses of data and models produce information that feeds into the decision-making process, where management objectives are taken into account and where another source of uncertainty occurs due to inclusion of difficult-to-quantify tradeoffs. The translation of the regulatory policy into specific tactical regulations is usually imperfect and may not achieve the desired goal. Furthermore, due to uncertain human behavior, the outcome of such specific tactical regulations is not clearly predictable because of the uncertainties of noncompliance and implementation error on the fishing grounds.

An additional source of uncertainty in fishery systems is the prospect of environmental, ecological, economic, social, and political change. In particular, our understanding of the

Fig. 4.1. Generalized scheme for flow of information in a fishery system, along with sources of uncertainty (shown as ellipses).

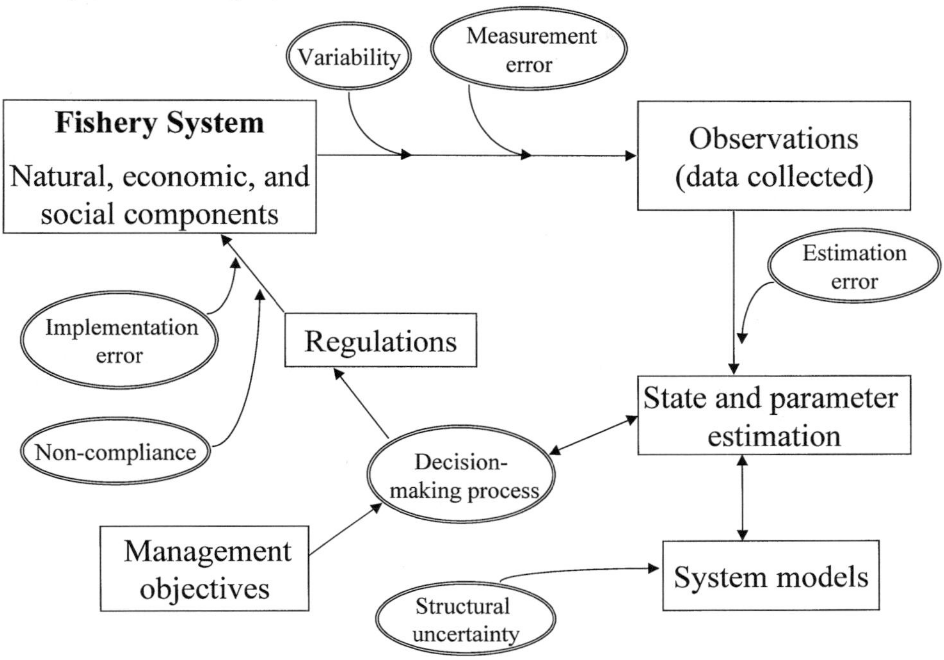

factors structuring these systems depends on observations of past behavior in response to changes in their various components. Because the past may no longer offer a reliable guide to future responses in the presence of environmental or other forms of change, an increase in uncertainty must be anticipated and effectively integrated into the decision-making process.

We discuss these sources of uncertainty for each of the major components of fishery systems: physical, biological, and human. The main purpose of this discussion is to identify the nature and extent of uncertainties and to make it clear that problems in fisheries do not solely arise from one source of uncertainty. Furthermore, this examination of sources of uncertainty helps identify some principles for dealing effectively with them, which will be described in Chapter 5.

Physical components

Measurement error

There is always some uncertainty associated with any oceanic measurement. However, for most physical measurements, such errors are small relative to the measured value. Water properties at a specific location can, in general, be measured with great precision and accuracy. Thus, temperature, for example, can be measured very precisely and accurately to ±0.001°C, while density can be measured to 0.001 kg/m^3. Other oceanographic parameters, such as those related to currents, cannot be measured as accurately or precisely.

However, a single accurate measurement of temperature does not tell us much about the ocean outside of some relatively small radius. It is therefore quite common to have insufficient data to adequately map out water properties over large spatial scales, such as tens or hundreds of kilometres. It is often necessary to interpolate between measurements spread out in space. However, water properties are so heterogeneous that a map of these properties can be quite inaccurate, even though the individual point measurements are quite good. Therefore, the ocean is unlike the atmosphere, where wind, temperature, and pressure measurements made at a single point are often representative of air masses spread over hundreds of kilometres.

Structural uncertainty

In making predictions of the physical ocean, uncertainty also arises because of the character of our physical models. These models are usually initialized with data that are relevant for a limited period (days to weeks). The models drift away from their initialization because they cannot adequately represent the small-scale non-linear processes that control circulation and water properties. Nonetheless, oceanic models have improved greatly in the past ten years and are increasingly useful in application to biological problems (Kendall et al. 1996). Over very short time scales (minutes to hours), we may have reasonable precision and accuracy over very large spatial scales. For longer time scales, the spatial envelope of a given level of precision may shrink considerably. The "conveyor-belt system" of ocean circulation discussed in Chapter 3 is a good example of the uncertainty that arises over longer time scales due to the structural complexity of oceanographic processes and their resulting dynamics. Our modelling capability is most uncertain at small spatial scales and long time scales. Failure to adequately represent non-linear processes, friction, and mixing is behind both the temporal and spatial scale limitations (Kowalik and Murty 1993; Lynch and Davies 1995).

Nonetheless, the major physical oceanographic processes are well enough described by existing models that they are reasonably successful at forecasting the qualitative character of the ocean, as averaged over larger space and shorter time scales (Mann and Lazier 1996). For instance, on the west coast of Canada, partially successful forecasts have been made several months in advance of when a general region-wide condition of unusually warm water will prevail as a result of an El Nino-Southern Oscillation event. This is important for fish because such warm waters are usually relatively poor in food for Canadian pelagic fish and rich in predators of those fish, leading to reductions in survival rates.

Biological components

Measurement error

Our uncertainty is greater for almost all biological quantities than for physical variables such as temperature and salinity. We can be quite confident about the accuracy of specific measurements made about individual fish (e.g., length and weight), but accuracy and precision are typically poorer for estimates of abundance or survival rates for populations of fish in the ocean. The resulting uncertainty results from the greater difficulty inherent in making biological measurements. Direct sampling using nets is usually necessary, with resulting limitations in resolution, potential biases associated with technique, and inefficiencies of effort due to high costs (very few measurements can be made). Further complicating the sampling and increasing the uncertainty is the well-known patchiness of marine organisms. An estimate of fish abundance in a small region is almost impossible to extrapolate without additional corroborative information.

As noted in Chapter 2, estimates of population abundance of exploited marine species are typically derived using methods such as sequential population analysis on age-structured catch data or by fishery-independent monitoring programs. Abundance estimates for the most recent years of such analyses are more uncertain. In many instances, research vessel survey indices are used to calibrate the sequential population analysis. Sampling error in the research vessel survey index therefore translates into imprecise total population estimates derived by sequential population analysis. Measurement errors on fish abundances are thus typically large. It is common for the confidence intervals on abundance measurements of this type to range from one half to double the mean estimated abundance (Sissenwine et al. 1983). Such imprecision means that it can be difficult to identify in a timely manner either increasing or decreasing trends in abundance over time, let alone to identify causes of observed trends.

Nevertheless, for many major fish stocks, management agencies have found an appropriate balance between costs of sampling and numbers of samples. It is commonplace to find sampling programs that are both extensive (spatially) and intensive (numbers of samples, as well as frequency of sampling within fishing seasons). An example of the latter is the weekly or even daily update of estimates of abundance and composition of harvests in mixed-stock salmon fisheries on the Pacific coast (Woodey 1987). As a result of this process and despite the sources of imprecision and bias described above, estimates of abundance have been sufficiently good to allow within-season adjustment of harvesting regulations to permit meeting spawning abundance targets. In some cases, these procedures have been able to rebuild abundances, e.g., the Horsefly River sockeye salmon stock on the Fraser River, B.C. Furthermore, for marine fish populations, we can generally forecast reasonably well their mean behavior and trends, at least over the short term. These mean, qualitative trends require changes in management strategies for changing productivity in ways that harder-to-forecast, shorter term, random fluctuations will not.

Structural uncertainty

Uncertainty in biological measurements contributes to limitations on our knowledge of the structure of biological processes in the ocean. For biological systems there exists nothing comparable to the well-accepted dynamical equations of physical oceanography. For instance, there is often considerable uncertainty about the shape of relationships between variables such as spawner abundance and reproductive success and about the existence, or relative importance, of links between environmental measures and fish survival rate. Indeed the characteristically high variability in recruitment has led some scientists to conclude (erroneously) that there is *no* relationship between adult population size and resulting recruitment (for a recent example, see Gilbert 1997). Clearly such a stance will very likely lead to risk-prone harvesting policies (such as the reduction of spawning stocks to extremely low levels). Aside from the problems of natural variability, short data series, and measurement error, another reason for the uncertain structure of models of fish populations is that biological systems are characterized by lags, non-linearities, age structure, and threshold phenomena, as well as considerable variation among individual fish within a population, let alone an aquatic community. This structural complexity creates complex dynamic behavior, where several different future levels of abundance are possible. Furthermore, information on interspecific interactions (predator–prey and competitive interactions) and the relationship between organisms and their environment (habitat) can also be difficult and expensive to obtain, making it hard to properly parameterize relationships in biological models.

The diversity of marine biological systems also adds to this structural and dynamic complexity. For instance, in coastal upwelling zones, species such as sardines and anchovies perform similar functional roles in terms of energy flow in the communities and may compensate to some extent for changes in abundance of the other species. Thus, disturbances that reduce abundance of one species may have less effect on the overall functioning of the community than might be expected. In this case the biological diversity of a community can lead to unexpected responses and can enhance the community's resilience to disturbance. In stark contrast to this, there are other species that perform such a dominant role, that if their abundance is changed due to natural or human causes, they can greatly affect the biological structure of their communities. Sea otters are one example; they eat sea urchins, which in turn eat kelp, amidst which some pelagic fishes spawn or live year-round. Reductions in abundance of sea otters can lead to unexpected, counterintuitive changes (increase in sea urchins and destruction of kelp beds) (Estes et al. 1978). A similar major effect of sea urchins was found on the Atlantic coast (Johnson and Mann 1988) when a disease caused their rapid decrease in abundance, while kelp beds, which they had previously grazed down, recovered. Thus, while evidence is accumulating that biological diversity may enhance a community's persistence (Naeem et al. 1994), such diversity also creates considerable uncertainty on the part of scientists and managers about the structure of communities, as well as their dynamics.

Nevertheless, there are numerous useful models of fish populations and communities that serve as forecasting tools and these are continually being refined and used effectively in management of fisheries (Ricker 1975; Gulland 1983; Hilborn and Walters 1992). For marine fishes, they can provide reasonably good short-term forecasts (the next 1 to 3 years) of population dynamics and responses to management regulations. However, longer-term forecasts are not as reliable.

A more insidious form of structural uncertainty in recruitment processes can be identified in cases where there are "multiple stable states" for abundances. Peterman (1977) and Steele and Henderson (1984) showed that recruitment dynamics characterized by depensatory processes (high mortality rate at low abundance due to predation) can lead to sudden changes in abundance. Such changes create serious consequences for both the resource and for the human communities dependent on it. If the depensatory character of the recruitment dynamics is unrecognized, depressing the adult population below some threshold level will result in a sudden collapse in recruitment with little or no warning. Even if the depensatory nature of the system is recognized, there may be considerable uncertainty about the position of the lower threshold level, thus creating extra risk.

Systematic errors that arise because of misspecification of key parameters or because of incomplete model structures (e.g., ignoring predation effects) also have critical effects on forecasts. Multispecies extensions of sequential population analysis have been implemented in intensively studied regions such as the North Sea and it has been found that the mortality rates attributable to predation are typically higher than the natural mortality rates employed in most single-species analyses for the younger age-classes. In such cases, single-species analyses produce an underestimate of the abundance of these age groups. In general, specification of the natural mortality rate of exploited marine species is difficult because undisturbed populations subject only to natural sources of loss are now uncommon and direct measurement is difficult or impossible. As a result, most analyses use a constant natural mortality rate and repeat calculations for different assumed rates. However, the choice of the natural mortality rate used in analytical methods such as sequential population analysis has a dramatic impact on the estimates of total population size and fishing mortality (Lapointe et al. 1989); populations can be overestimated if the estimated natural mortality rate is too high.

In a broader ecosystem context, changes in fundamental production characteristics (role of various species in the community) will necessarily affect the sustainable levels of exploitation for a fishery system as a whole (Chapter 3). Further, to the extent that environmental change induces fundamental structural changes in marine ecosystems (e.g., through changes in species distribution patterns, species composition, food web structure, or in energetic pathways), current understanding of population, community, and ecosystem processes may be insufficient to predict future response patterns.

Human systems: the economic and socio-political components

Measurement error

Issues related to measurement error are also important factors within economic and socio-political subsystems of fisheries. Morgenstern (1963, p. 7) notes that "...at least all sources of error that occur in the natural sciences also occur in the social sciences ... the statistical problems of the social sciences cannot possibly be less serious than those of the natural sciences. Indeed, the situation is far more serious ..." For instance, it may be difficult to extrapolate observed behaviors across entire groups of fishermen, government decision makers, or shoreworkers based on a few samples. Furthermore, interview coverage will be limited by the funds available to conduct, analyze, and interpret information on prices and costs, thereby reducing precision of the estimates for these basic economic variables. In addition, biases may occur because harvesters and others involved in fishery systems may misreport catch data inadvertently or purposely. They may also be unwilling to provide information on important economic measures. In general, administrative data gathered for management purposes are not appropriately organized to serve as observations for purposes of research aimed at understanding the relationships underlying social and economic systems.

Measurement error comes from two sources, one where some variable is not measured at all and the other where the variable is measured but in some imprecise or biased manner. Relative to the biological and physical components of fisheries, many fewer measures of economic and especially socio-political components tend to be made. For instance, there is currently a lack of data on several key variables that are important for evaluating what social and economic effects proposed regulations will have. We lack sufficient data on where license holders live, what strategies multiple license holders use (e.g., emphasizing fishing on salmon as opposed to herring), and how the predominant "share system" between vessels owners and crew is shifting in response to new policies. In addition, we lack data on fishing effort and harvests that do not pass through the "official" market.

In other cases, certain data are collected but are not sufficiently analyzed to provide clear indications of effects of new management regulations. Examples include the views of fishermen on how much owner–operators should be favored in granting licenses over other types of owners, the frequency of different categories of multiple license holders, and levels and types of debt held by fishermen. When such data are collected, they suffer from relatively large measurement error due to small sample sizes and the diversity of human behaviors over space, time, and individuals. Similarly, although management policies can greatly affect the economic viability and social stability of coastal communities, it is difficult to estimate how dependent communities are on fishing. Employment statistics by occupation do not accurately reflect the diversity and seasonality of occupations and thus are prone to error. The complete lack of certain kinds of economic and social data, combined with limitations on other data, also makes it very difficult to measure accurately the complex and multifaceted impacts of humans on the ecological system. Intangible assets such as social or cultural capital, as well as more tangible but still priceless natural capital are widely recognized to be crucial attributes of the ecosystem we study; however, no data are systematically collected on them and even in the case of natural capital, monitoring efforts are sporadic at best. Full assessment of the environmental risks arising from contemplated action is therefore rarely possible (see also Box 4.2 below).

Structural uncertainty in social and economic systems

Structural uncertainty affects efforts to model economic and social components of fishery systems. For example, the shape of the production function (e.g., relating the eventual harvest to fishing effort and other inputs) or the demand curve (relating demand to price) cannot be known with accuracy. In many instances, economic models are integrated with biological submodels and the economic outputs are subject to all of the uncertainties in the underlying biological structure.

Lack of any empirical foundation from which to anticipate the behavioral response of individuals or groups to unanticipated events or to policy interventions is a particularly unfortunate feature of much of fisheries policy analysis. It not only generates the same sorts of irreducible uncertainty as arises from the complexity of natural systems, but leads as well to reliance on predictions from highly simplified theories of individual behavior. For instance, disagreements among economists about the incentives that drive the behavior of humans involved in fisheries (e.g., Pearse and Wilen 1979; Copes 1997) will persist undiminished until the question can be clarified to some extent through empirical testing.

A major source of structural uncertainty lies in the fact that, while there have been excellent "one time" studies of the socio-political structure and dynamics of Canadian fisheries (e.g., Marchak et al. 1987), they tend not to be incorporated into regular decision-making processes. Hence, processes that link the socio-political components of fishery systems to other components are not normally taken explicitly into account. More generally, Lindblom and Cohen (1979) underline the extent to which all knowledge and science are embedded in a complex social and political order about which research and analysis can never speak authoritatively. Wynne (1992) argues for recognition of ignorance and indeterminacy in human systems as generating new kinds of inevitable uncertainty, not simply broader bands of imprecision that might be narrowed through further research.

Responses to uncertainty

As noted, uncertainty creates risks. It is instructive to examine the three general ways in which people respond to the presence of uncertainty. The first is to ignore uncertainties. This involves decision makers (be they government managers, scientists, members of fishing industry, or individual fishermen) taking the best estimates of all parameters and state variables, assuming that they are correct, and acting accordingly. This tactic reflects the common approach where scientists are pushed to give a single answer, rather than a range of possible answers, which would more appropriately reflect the inherent uncertainty. This tactic also reflects situations where decision makers do not consider the scientific information, either because it is lacking or because it is not well-understood.

A second approach is to take uncertainties into account qualitatively. This approach manifests itself in four different ways. (*a*) Use uncertainty as a reason to maintain the status quo and not to deal with the concern at hand. For instance, argue against the reduction of emissions of greenhouse gases in the face of uncertainty about climate change. (*b*) Use uncertainty to justify a strong response to the concern at hand by applying the precautionary principle. For example, don't allow any dumping of wastes in the North Sea. (*c*) Use uncertainty to justify a moderate response to the concern at hand, i.e., take a guarded or precautionary approach. For instance, allow harvesting but use some arbitrarily set safety margin, for example, 25–50% reduction in harvesting quotas on groundfish in B.C. (*d*) Use uncertainty to justify optimism, i.e., adopt an aggressive approach to intervention in natural systems. For instance, select the upper bound on an estimate of stock abundance when choosing a catch quota.

A third approach is to take uncertainties into account quantitatively through formal methods such as risk assessment or decision analysis. These methods often involve assigning a probability to each potential state of nature (and its consequences) and then using these probabilities as "weightings" on the possible outcomes to calculate the average, "expected" result, as well as various other measures of outcome relevant to risks. Risk assessments and decision analyses can be carried out taking into account whatever management objective and whatever degree of risk aversion managers desire.

None of the above analytical approaches is the best option in all situations and some are not appropriate in any situation. In particular, it should be noted that these methods of formal decision analysis may not be appropriate in cases where there is considerable uncertainty that cannot be quantified or even categorized and where risks are large. This is because decision analysis requires the analyst to specify a range of hypothesized values of parameters or range of relationships, which are then weighted in some way, usually based on field data. However, these methods have considerable potential for use in fishery systems; in contemporary fisheries management, efforts are being made to structure decision-making processes to deal more explicitly and effectively with uncertainty and risk. Therefore, in the next sections we explore briefly the ways in which lack of knowledge about natural and human systems might result in errors created or transmitted through management processes.

For convenience, we organize the discussion below around a decision cycle which begins with risk assessment (identification, characterization, and analysis of risks and consequences, with some assessment and evaluation of consequences and tradeoffs among them) and then move on to risk management (identification and appraisal of possible actions and selection of a preferred action from among them) and finally to implementation of the preferred action and monitoring of the results.

In particular it is important to note that in human systems and certainly in fisheries management, action is usually taken in an organizational context, implemented through the concerted efforts of many individuals. These individuals may be part of a single agency or organization with well-established executive authority; more likely they will be spread over many organizations, contract agencies, co-management groups, corporate enterprises, vessel crews, and individual actors. In any case, unlike physical or biological systems, these entities pursue their own purposes and implementation of decisions can only be accomplished through a chain of action which enlists the cooperation of all the individuals involved from executive agencies out to vessels on the sea.

Risk assessment: appraising the effect of uncertainty

Risk perceptions

When scientific evidence and decision analysis are mobilized in support of a group decision, it is necessary to recognize that data, uncertainties, and risks will be differently perceived and interpreted by different individuals (Fischhoff et al. 1978; Kahneman and

Tversky 1979). Typically, public perceptions of risk tend to differ from expert rankings or estimates of risks calculated from data (Slovic 1987). Thus, some interactive process of reconciliation and synthesis of beliefs and perceptions may be necessary, with decision analysis perhaps employed to make clear the implications of alternative views (Lindblom and Cohen 1979; Raiffa 1982). Further, this interactive process must take into account, in effect, uncertainties surrounding the realization of decisions themselves. For example, although the information emerging from decision analysis may lead to development of consensus around a desired harvest level in a fisheries management setting, that harvest level is something no individual or group can directly control. Thus, uncertainty must be addressed not only analytically but also procedurally, through interactive processes which seek a synthesis of disparate views and beliefs about the nature and state of the system, reconciliation of perceptions of uncertainties and risks which may not correspond to statistical evidence, and consensus around the appropriate responses to risks, given differing distributions of costs and benefits associated with each possible outcome. In other words, risk assessment is not just a "technical" matter of making calculations, but is very much a process involving human interactions. It is inextricably linked to judgments about appropriate responses or in other words to risk management, the subject of a later section.

Valuation problems

Ideally, before some activity such as harvesting of fish or disturbance of habitat is pursued, the various ramifications of the activity should be fully evaluated to account for a wide range of potential effects. For example, the cost of a hydroelectric dam should in principle include not only the cost of materials and labor for construction and operation, but also, for example, some measure of the impacts on salmon stocks, as well as the ecological effects of flooding habitat, and changes in the carbon budget as new areas are flooded and decompose, giving off carbon dioxide. Unfortunately, both conceptual and logistical problems make it difficult if not impossible to measure such impacts. Resource economists have developed methods for estimating the non-market value of such environmental effects in order to permit decision makers to incorporate at least some of them more fully into decisions, but the attempts are controversial (see Box 4.2).

Box 4.2. Valuing Environmental Risks

As Knetsch (1993) observes, "These are interesting times for resource and environmental valuation ... While the enthusiasm for valuations has perhaps never been stronger, the evidence that we may indeed be getting misleading answers to the wrong question is pervasive."

On one hand, the demand for valuation estimates has grown, particularly in the U.S., not just in project appraisal and policy analysis, but from litigation and negotiation to establish compensation for environmental harm. As one response, in the aftermath of the Exxon Valdez accident, Congress assigned responsibility for development of a method for assessment of environmental costs to the National Oceanic and Atmospheric Administration, which developed its "Blue Ribbon Panel Guidelines" (NOAA 1992) which now serve widely as a handbook for practitioners. More generally, researchers such as Costanza et al. (1997) insist on the necessity of valuing ecological services in an ever more comprehensive fashion, seeking estimates of the value of ecological systems comparable in scope to economic indicators such as gross domestic product.

On the other hand, a variety of critics point to empirical evidence of serious limitations to such efforts. Two key concerns are cited, beyond the perennial difficulties associated with the discounting of future benefits or costs, whether environmental or otherwise. The first is the apparent systematic disparity between valuations of gains or losses, signaled by consistent discrepancies between estimates of willingness to pay (WTP) for potential gains and willingness to accept compensation (WTA) for potential losses. The second concern is growing evidence that the overwhelmingly most popular method for assigning monetary values, the contingent valuation method (CVM), yields estimates which are unlikely to be the measures of economic values they are intended to be (Knetsch 1993).

Faced with this dilemma, some observers suggest procedural alternatives. Blore (1996) suggests that "Contingent valuation may be valuable to decision makers not because it gives good answers to questions about the valuation of the environment, but because it has merits as a process within a policy-making framework." These merits include forcing policy-makers and the public to be imaginative about options in a context which builds in the problems of risk and concepts of subjective risk. Gregory et al. (1996) suggest processes that "offer promise of yielding information superior to the valuation numbers provided by more traditional methods." One would be to focus on the concerns of a small group of people selected to be representative of the key interests potentially affected by a proposed action; a second would rely on a small group of people convened as a "values jury"; the third centers on the derivation of a "damage schedule" providing scaled rankings of the relative importance (rather than impossible-to-measure absolute values) of various environmental harms.

As Pimm (1997) notes, however, there are also many critics (e.g., Sagoff 1997) of the whole enterprise of environmental valuation who believe that it presses policy formation toward a misguided consequential approach based on calculation in a situation where only an ethical and rules-based approach can offer a legitimate foundation for decision. With or without explicit valuation efforts, this literature again underlines the facts that decisions about environmental harm or the value of ecological services cannot be considered technical matters or confined to expert judgment and that there is considerable uncertainty about consequences of actions.

Specification errors in policy formation

Measurement errors and structural uncertainties or broader indeterminacy in social and economic systems may lead to a variety of errors in policy making, generally referred to as "specification errors." For instance, an incorrect understanding of the social structure in a fishery (based on the structural uncertainties described above) may lead to management measures that produce catch allocations or changes in resource access that are contrary to societal objectives. An example might be the measures to achieve reduction of fleets and fishing effort on Canada's west and east coasts, which, apparently unintentionally, have led to a reallocation of quota and fishing rights away from smaller communities toward corporate and (or) urban interests. It may be argued that this result is unlikely to achieve the balance of objectives desired by Canadian society.

Measurement errors may result in an inability to achieve exactly target levels of yield or exploitation. For example, consider catch quotas, which are often set based on a preseason forecast of abundance. To the extent that the estimated preseason abundance differs from actual abundance, the quota will be misspecified with respect to the management objective. The higher the uncertainty in the estimated abundance, the higher the probability that the management objective will not be met. Rosenberg and Brault (1993) show that this problem is particularly severe when the error in estimation is linked to the abundance of the stock. In other situations, errors in estimation of catch will result in errors in forecasting and hence in implementation.

Implementation error in fishery management systems

In addition to the uncertainties described above with respect to the social and economic components of fishery systems, we can also identify uncertainties and consequent "implementation errors" associated with implementing management programs or with the human response to regulatory controls (Rosenberg and Restrepo 1994).

In conventional regulatory approaches to management, such implementation errors may be due to a lack of understanding on the part of managers of the behavior of fishermen and a consequent failure to incorporate an allowance for such behavior in designing management measures. This can be reflected in many ways. Managers in a salmon fishery may incorrectly predict the "behavioral response" of fishermen to a fishery opening in a particular location, possibly leading to under- or over-harvesting. Changes in policy or in management systems may produce uncertain responses by fishermen, including the possibility of resistance among harvesters (non-compliance with regulations). This may appear through illegal activity of various sorts, including misreporting of catches and other fishery-based information, resulting in a failure to see policy objectives fulfilled, and a general degradation of the information base used to make management decisions in the future. For example, Kearney (1984) reported large underreporting of catch by holders of seine quotas in the Bay of Fundy herring fishery. Similarly, in New England, the imposition of catch quotas designed to promote stock recovery in the initial years following the implementation of extended jurisdiction was extremely unpopular among harvesters. Widespread civil disobedience, misreporting of catches (both species identity and location caught), and cheating were documented (Peterson and Smith 1982).

In addition to errors produced in the immediate fishery management process, implementation errors can be expected at a system-wide level. This is due to the fact that implementation involves an extraordinarily long chain of action through legislation (possibly), regulatory initiative, administrative measures, out into related agencies and field offices, into remote regions, and finally into the conduct of individuals exercising their own individual discretion in making personal decisions from moment to moment on fishing vessels (Calista 1994; Dobell 1998). It is hardly surprising that a policy goal formulated in a central agency may be reinterpreted many times in the light of varying and changing circumstances at different points in the system. The development of fishery policy involves high stakes, possibly risking the irreversible loss of an irreplaceable component of a common human heritage, productive fish stocks. It also involves the livelihood of the many individuals involved, who must be expected to pursue their own objectives and their own interpretation of responsible behavior, not the abstract instruction of some distant office. We have discussed various quantitative methods to assess risks and balance these contending interests in support of policy formulation. However, the procedures by which decisions are made, are also crucial, as discussed below.

The need to achieve the commitment of all those involved in the implementation process has only recently been fully recognized in theory. It leads to emphasis on institutional design in addition to policy formation. In particular, it strongly suggests that fishermen should be directly involved in the development of management strategies at an early stage. It further indicates the need for well-defined rights and incentive structures designed to foster stewardship and conservation of the resource among user groups. This emphasis on the importance of participatory mechanisms to build consensus and commitment and on the design of institutions to assure that the structure of incentive systems promotes, rather than works against, individual compliance with agreed regulations and plans, provides the focus and foundation for our later discussion of principles and strategies for sustainable fisheries.

It is important to emphasize the need for unforced voluntary commitment of individuals if policy intentions are to be realized in the uncertain, variable, and changing setting which we have identified as the fundamental characteristic of fisheries systems. Beyond concerns for the clarity of mandates and policies, the effectiveness of sanctions, or the accuracy of forecasts, there is the central importance of the perceived legitimacy of the process by which policies and regulations are developed. If harvest levels are not perceived as legitimately derived or if government-set quotas are not considered appropriate by fishermen, then full compliance with rules and realization of target catches is unlikely. So also is achievement of a resilient, self-regulating system in which action can be adapted to new information on a continuing basis to achieve agreed upon objectives in changing circumstances.

Compensation mechanisms for individual risks

Within the framework of human institutions are many provisions that alter the magnitude of the consequences facing individuals in risky situations. In the face of pervasive uncertainty about both the underlying fish stocks and the external competitive setting, it is difficult for individual participants in the system to separate short-term fluctuations from longer-term trends, whether in catches, markets, or in economic outlook generally. Their decisions whether to increase or decrease their investment and their involvement (financial, personal, or social) in fishing activities will depend on their assessment of uncertain future outcomes. This assessment includes consideration of the consequences, or "payoffs," to them individually associated with these outcomes. Their decisions also ought to take full account of the compensation mechanisms (private or social insurance, subsidies, other compensation payments) and transfer programs (Employment Insurance, special assistance for mobility or retraining, etc.) that in some cases are contingent on the outcomes.

The design of such compensation programs is critical, at times involving fundamentally unresolvable balancing acts. If short-term downturns or even collapse in stocks are signals of enduring change, then labor and capital of all kinds (financial, physical, human, and social) should presumably be moved out of harvesting, processing, and other associated activities. There is no sense sustaining such labor and capital "on the sidelines" while waiting in vain for recovery of fish stocks. However, this brings up the difficult-to-estimate value of losses caused by people having to abandon their home or way of life. On the other hand, if these observed developments are merely short-term fluctuations, then social provisions to pool the risks and share the burden of temporary adverse contingencies should be justifiable and compensation or transfer mechanisms help those affected to bridge the bad times. (Of course, there may still be a debate about whether government support is required or whether such "bridging" mechanisms should be paid for through a self-insurance fund that is paid into by future beneficiaries during periods when economic revenues are large, as is commonly done in agriculture.)

On the positive side, the presence of such compensation mechanisms has the effect of cushioning the negative economic consequences of downturns and thus reducing risks (both perceived and real) for corporations and individuals. This has the very important implication of reducing the pressure for immediate harvests, since fishermen and plant workers are not forced to obtain all their income from such harvests. If fish stocks are low, so that harvests are poor, there may be greater use of compensation mechanisms and less threat of overharvesting. However, on the negative side, compensation payments to firms may have the effect of encouraging a "ratcheting up" (increasing harvesting capacity more in good times than decreasing in bad times) of investment and harvesting capacity over many years, even in the face of reduced catches and evidence of long-term stock declines (Caddy and Gulland 1983). The likely prospect of compensation to firms and fishermen if things go wrong may also work against appropriate precaution in their individual decisions. In the case of labor, compensation programs to smooth over interruptions have a mixed result. By deterring movement of people out of a fishery, they tend to maintain pressure by fishermen on managers to harvest a declining or still depleted resource, with consequent increased biological risks of severe stock depletion. However, they also maintain a "critical mass" of people in coastal communities (by reducing out-migration). Finally, while compensation mechanisms have a mixed impact on the various aspects of sustainability in fisheries, it is certainly true that an adequate social support network, together with methods for pooling risks and sharing costs, are essential to allow for adjustment of humans to changing circumstances in fisheries and in coastal communities. The various advantages and disadvantages of compensation programs should thus be considered very seriously in terms of their impact on biological, economic, and social risks.

Determining the balance between "smoothing" out fluctuations in economic gains and encouraging change is a matter for judgment, not easily made and always highly contentious. Also, one must take into account the potential irreversibility of actions, not only for the resource but also for resource-dependent communities in which the whole network of community connections, personal networks, and social capital cannot be easily recreated elsewhere (Ommer 1994; OECD 1981).

The debates on this issue of compensation programs and social support mechanisms are too extensive to elaborate further here. But it is important to remember that however compensation payments and transfer programs are viewed, their presence has significant implications for the effectiveness of policy initiatives, the likelihood that policy intentions will be faithfully implemented, and the appropriateness of future pressure placed on fish stocks.

Risk management: decision-making in the presence of risks

We have seen that uncertainty is pervasive in fishery systems. Gaps in knowledge of the component systems or uncertainties in understanding their structures and their states may be translated into errors in policy formation and in implementation, as just described. How can we constructively deal with uncertainty and the attendant risks? This is the so-called risk-management step, as opposed to the risk assessment or risk analysis step (where estimates of risks are made based on the uncertainties and their consequences). Here, we argue that to manage risks sensibly, we require some explicit framework to determine the implications of uncertainty in fishery systems and to guide management decisions.

Experience in dealing with risks explicitly has been gained by those making business decisions, such as industry investing in new processing plants and by stock assessment biologists who work with decision makers, among others. There is a growing literature on methods to represent uncertainties in environmental, biological, and even some economic components (e.g., Helton and Burmaster 1996; Punt and Hilborn 1997; National Academy of Sciences 1998). However, in most cases, there is an unfortunate lack of documentation with regard to how risks have been dealt with in practical fishery situations. In the case of government management agencies, scientists pass on information about uncertainties to decision makers to varying degrees and then the decision makers use that information in some way to make tradeoffs and evaluate options. Because of the lack of documentation, we can report little on what has been done in fisheries risk management, except for some cases of estimating stock abundance and evaluating management options. This lack of "transparency" of the process is one reason why there is a debate among some scientists about what information is used by decision makers, how seriously uncertainties are acknowledged, and how fisheries management could be improved (Stephenson and Lane 1995; Hutchings et al. 1997; Doubleday et al. 1997).

This section addresses the challenge of developing an appropriate response to uncertainties, leading to recommended management measures and harvest strategies. It then considers uncertainties associated with implementing those recommended strategies and institutional arrangements and other measures designed to assure compliance with the implied restrictions on individual action.

Fish stock assessment, uncertainties, and implications for management

Ferson and Ginzberg (1997) argue that the two principal sources of uncertainty (inherent variability and incomplete knowledge) must be treated in fundamentally different ways. In instances where we can describe the inherent chance of obtaining a particular result, probabilistic methods can be applied. These are the methods mainly used in stock assessment to account for uncertainties and risk when evaluating management options. In other cases, where incomplete knowledge forces people to make various assumptions or choices of inputs to the analysis, we may only be able to delimit the outcomes by using the extremes in a sensitivity analysis.

Applications of formal risk assessment and decision analysis, which by definition include probabilistic approaches, are becoming increasingly common in evaluations of fish stock abundance and management options (e.g., Walters 1981; Brown and Patil 1986; Linder et al. 1987; Bergh and Butterworth 1987; Fogarty et al. 1992, 1997; Francis 1992; Restrepo et al. 1992; Rosenberg and Restrepo 1994; Hilborn and Peterman 1996; Huppert 1996; Kirkwood and Smith 1996; Punt and Hilborn 1997). The symposium volumes edited by Shepherd (1991) and Smith et al. (1993) contain excellent overviews of the applications of risk analyses to problems in fishery management. Many expressions of risk are possible depending on the choice of fishery management objectives. To provide a sense of the types of risk analyses that have been conducted in fisheries, we provide several examples below. These vignettes illustrate both the types of uncertainty and kinds of results that are possible.

A common problem in fisheries management is to estimate the tradeoff between harvest now and future risk for each of several management options. Restrepo et al. (1992) examined this type of problem for the northern cod off Newfoundland based on data up until 1990 and estimated that, for example, a total allowable catch of 210 thousand tonnes would imply about a 15% chance of exceeding the target fishing mortality rate — which happened to be the rate in the previous year (Fig. 4.2). They also estimated that an increase in total allowable catch to 220 thousand tonnes would result in a probability of 40% of exceeding the target fishing mortality. Thus, in this case, a relatively small increase in TAC (total allowable catch) from 210 to 220 thousand tonnes resulted in almost a tripling of the probability

Fig. 4.2. Cumulative probability (S-shaped curve) that the status quo fishing mortality rate target will be exceeded under different total allowable catch (TAC) levels, and the foregone yield (curve shown with its 95% confidence interval indicated by dashed lines) if a given TAC is used (reprinted from Restrepo et al. 1992).

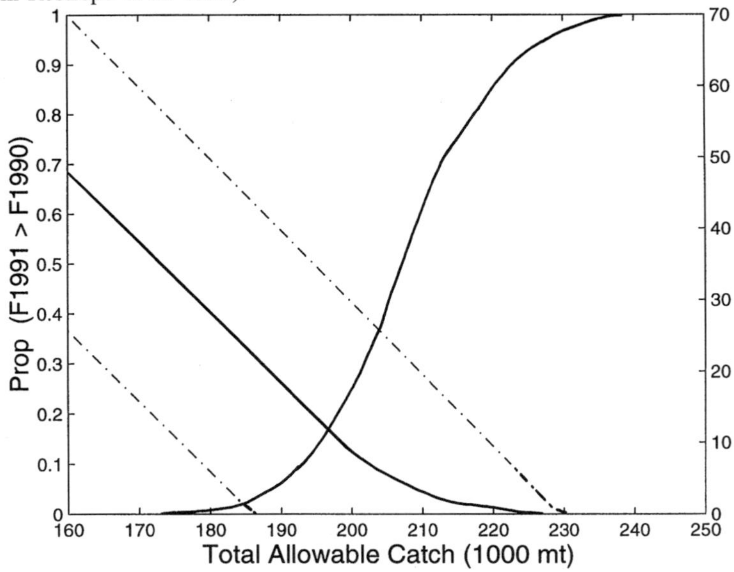

of exceeding the target fishing mortality. Such a presentation allows managers to make a direct evaluation of the tradeoff between this particular type of risk and harvest. Of course, this specific analysis says nothing about the probability of resource collapse for different harvest rates, but such analyses have been done for other stocks.

Another challenge is to describe the uncertainty in estimates of fish stock abundance in any year, given the numerous sources of uncertainty. It is possible to do this for sequential population analysis (SPA), the most common method used to reconstruct past abundances of fish stocks. In such analyses, the SPA is repeated a large number of times to specify the probability of obtaining a particular estimate of population size or fishing mortality. Figure 4.3 shows a population trajectory (for abundance of age-5-and-above swordfish in the Northwest Atlantic) with 95% confidence intervals.

In the absence of full information on fish stock dynamics, many fisheries scientists and managers who wish to account for uncertainties and risks have suggested that a safety margin or adjustment be used to reduce harvest rates from some base case by an amount such as 20%, 30%, or even 50%. However, given all of the uncertainties in fish stock dynamics, it is not clear whether such arbitrary amounts are sufficient to protect a stock from overharvest, or too large (potential sustainable yield could be foregone in the latter case). Frederick and Peterman (1995) examined the implications of various uncertainties and found that the optimal adjustment varied with the life history characteristics of the population and the type of harvest policy (fixed escapement or fixed harvest rate). For example, for Atlantic menhaden, a pelagic schooling species, the optimal harvest rate became increasingly conservative as uncertainty in the stock abundance estimates increased (Fig. 4.4). Specifically, the optimal uncertainty adjustment was a 20% reduction in harvest rate if the stock abundance estimate was quite uncertain (coefficient of variation of 0.5), which generated a harvest 15% larger

Fig. 4.3. Trajectory of estimated population abundance for age 5-year-old and above swordfish in the Northwest Atlantic; 95% confidence intervals shown surrounding best estimates (Restrepo et al. 1992).

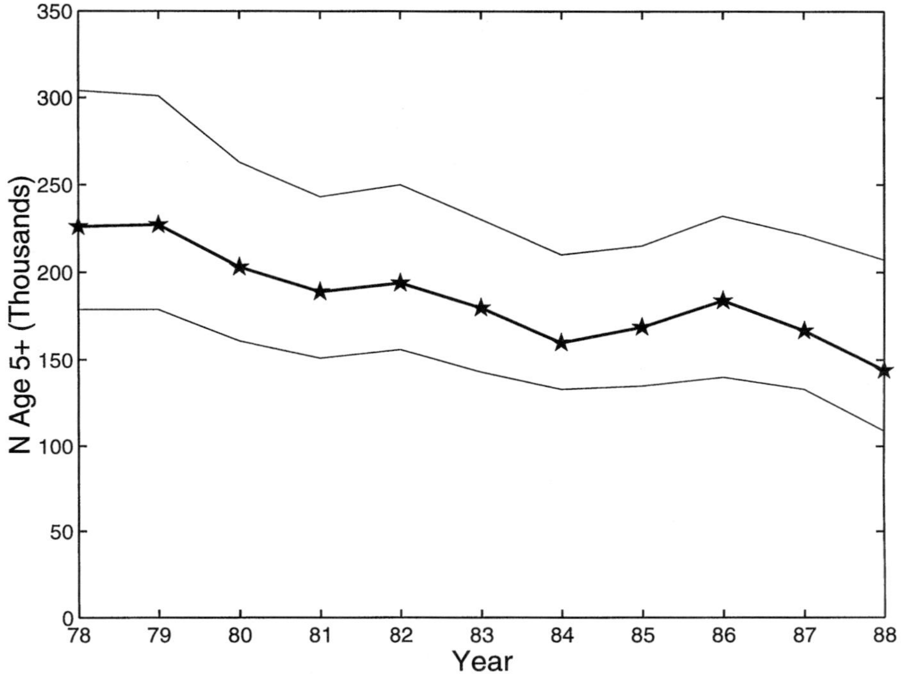

than if no adjustment for uncertainty had been made. However, this result applied only to Atlantic menhaden; the optimal adjustment is stock-specific. Frederick and Peterman (1995) concluded that uncertainty adjustments should be estimated for each individual species or stock and that managers should not use blanket prescriptions about conservative harvest levels to account for uncertainty because such arbitrary levels could be suboptimal. These adjustments to management regulations to reflect uncertainty are a method for developing precautionary "reference points" (see Box 4.3), which are becoming increasingly common in the management of fishery systems (Smith et al. 1993).

Another challenge facing fishery managers is that for a given stock and year, the limiting fishing mortality rate (one that they do not want to exceed) is uncertain, as is the probability of not exceeding it with given regulations. Fogarty et al. (1997) examined this type of situation for Georges Bank cod by simulating the effect of measurement error in population estimates and variability in recruitment on the estimated limiting fishing mortality rate. They then incorporated the variability in harvest rates and population size estimates derived from sequential population analysis to determine whether the estimated fishing mortality rates in a particular year (1993) exceeded the limiting level, given the uncertainty in each. Figure 4.5 illustrates the estimated probability distributions of fishing mortality rate in 1993 (F93) and the distribution of estimated limiting fishing mortality rates (Frep). Although not all sources of uncertainty were accounted for in this analysis, it is clear that the 1993 level of fishing mortality had a high probability of substantially exceeding the limiting level for

Box 4.3. Reference Points

A reference point is an estimated value derived through an agreed upon scientific procedure, which corresponds to a state of the resource and of the fishery and which can be used as a guide for fisheries management. The two key categories of reference points are limit reference points (LRP) and target reference points (TRP). Both reference points are expressed in terms of values of variables such as spawning stock biomass, harvest rate, or other measures. *Limit* reference points are conditions that are dangerous and should not be approached closely, let alone exceeded. In contrast, *target* reference points represent reasonable goals or objectives for management that have significantly less risk than LRPs (Caddy and Mahon 1995; Smith et al. 1993). These reference points can be considered precautionary to the extent that they are designed to maintain productive fish stocks in the face of the various pressures, uncertainties, and risks that characterize fisheries.

These concepts of reference points have been incorporated into Annex II of the 1995 UN Agreement on Conservation and Management of Straddling Fish Stocks and Highly Migratory Fish Stocks (UN 1995).

Precautionary reference points should be stock-specific to account, among other things,"for the reproductive capacity, the resilience of each stock, and the characteristics of fisheries exploiting the stock, as well as other sources of mortality and major sources of uncertainty" (UN 1995).

"Management strategies shall seek to maintain or restore populations of harvested stocks, and where necessary, associated or dependent species, at levels consistent with previously agreed upon precautionary reference points. Such reference points shall be used to trigger preagreed [upon] conservation and management action. Management strategies shall include measures which can be implemented when precautionary reference points are approached" (UN 1995).

For instance, some harvest rate that would generate a low abundance of a fish stock may be recognized as too risky because of the high probability of reduced recruitment, so this harvest rate would be a *limit* reference point, i.e., one to avoid. Fishery management strategies should ensure that the probability of crossing *limit* reference points is very low. If at some time, a stock has an unacceptable chance of crossing a limit reference point (e.g., harvest rate too high), conservation and management action should be initiated immediately to facilitate recovery of the stock. Even better would be a region associated with each LRP such that if current estimated conditions enter that region, preagreed upon actions would be taken immediately to avoid approaching the LRP any closer (Caddy and Mahon 1995).

In contrast, one plausible *target* reference point, or management goal, would be a lower harvest rate that would keep the stock at a moderately high abundance but would reduce the average yield in return for a lower risk. Fishery management strategies should ensure that *target* reference points are not exceeded on average. The poorer the data, the farther a target reference point should be from the limit reference point.

this stock. This means that the fishery was not likely to meet the objective that the limiting fishing mortality rate, Frep, was designed to meet.

The examples considered above deal with the issue of uncertainty in measurements concerning the status of fishery resources and the probability that management targets will be met. Relatively few researchers (e.g., Rosenberg and Brault 1993) have dealt with the

Fig. 4.4. Expected present economic value of cumulative catches of Atlantic menhaden for various uncertainty adjustments to the deterministic harvest rate (the rate that would be applied if only point estimates were used to choose management actions, thus ignoring uncertainties). The six curves represent results for different uncertainties, or coefficients of variation (CV), in annual estimates of the stock abundance. Solid circles on each curve indicate the optimal uncertainty adjustment (i.e., the one that maximizes the expected present value of catch) for each CV. Each uncertainty adjustment reflects an increase or decrease from the deterministic harvest rate (D) of 72%; e.g., applying an uncertainty adjustment of –20% resulted in an exploitation rate of 52%. Adapted from Frederick and Peterman (1995).

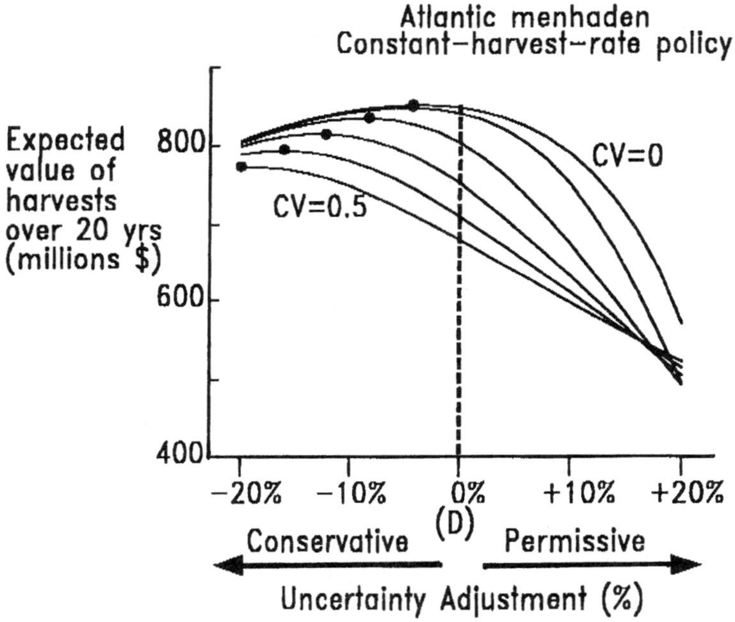

implementation error described earlier in this chapter, where the actual harvest rate achieved differs from the target level due to various factors in the dynamics of harvesters. Such analyses are a useful addition to the information on which decisions are based because they quantify the potential risks associated with deviation from desired harvest rates.

As discussed above, we often have large uncertainties about the structure of the underlying dynamic processes in fish populations and the human communities dependent on them. In fact, measurement error and natural variability in stock-recruitment data may be so great that several qualitatively different shapes of stock-recruitment curves may explain the data equally well, yet the different curves can make a substantial difference in recommended harvesting policies. To deal with such situations where there are qualitatively different states of nature possible in components of fishery systems, a formal decision analysis can be used (Clemen 1996). This is where various hypotheses or parameter values (i.e., "states of the system") are specified, their probabilities are estimated, and the consequences of these alternative states are explored. In the case noted above, there might be "structural uncertainty" about the form of the stock-recruitment curve that underlies some stock's dynamics. Alternative forms could be considered explicitly in a decision analysis to determine

Fig. 4.5. Probability distributions of the estimated "instantaneous" fishing mortality rate for 1993 (F93) and the limiting levels of fishing mortality (Frep) for Georges Bank cod (Fogarty et al. 1997). On the X-axis, a fishing mortality rate, F, of 0.5 results in a 39% harvest rate and F = 1 results in a 63% harvest rate.

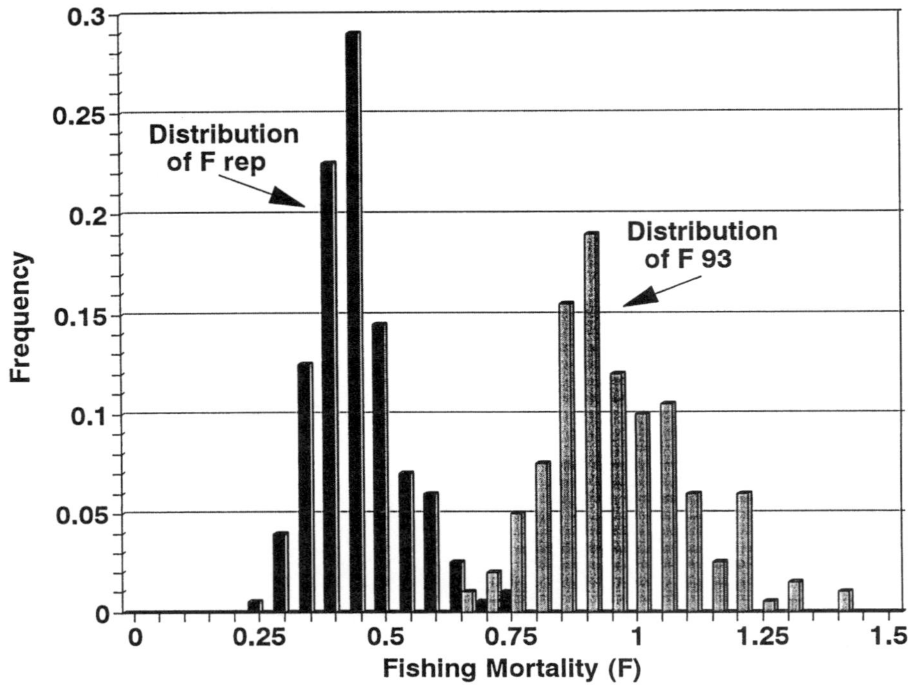

which management strategy would be optimal, given that uncertainty. Descriptions of this general approach in a fisheries context have been provided (Walters 1981; Rothschild and Heimbuch 1983; FAO 1995*c*; Punt and Hilborn 1997; Peterman et al. 1998*b*; Charles 1998).

Decision-making by industry in the presence of uncertainty

An example of the ramifications of uncertainty for industrial decision making is illustrated by the Western Australian rock lobster fishery system. The abundance of the post-larvae settling inshore (related to the strength of ocean currents, Phillips et al. 1994*a*) has been successfully related to the catch of recruiting lobsters 3–4 years later. Forecasts of the catch based on post-larval settlement indices have been used to guide investment decisions in the fishery and to anticipate changes in regulations that might be necessary to protect the stock (Phillips et al. 1994*b*). Uncertainty about how climate change might affect the relationship between major current systems and transport of lobster larvae in this case could lead to a breakdown in the prediction system used to monitor and manage this valuable fishery.

Social decisions and fisheries management policies in the presence of risks

Previous sections have outlined the many sources of uncertainty in fisheries systems and have sketched some quantitative methods that have been developed to take into account this uncertainty. These methods help to clarify considerably the choices to be made on harvest levels and fishing strategies. But these methods yield only clarification of tradeoffs and choices, not final answers. While they enable data subject to errors and uncertainties to be systematically analyzed in order to arrive at more summary information with estimated degrees of confidence, no technical method can replace human judgment in making decisions. Interpretations of the underlying information can be processed, through the methods mentioned above, to examine the implications of different harvest levels or management policies in terms of current yields and probabilities of future reduction or collapse in stocks. The inevitable balancing of risks — of estimated current benefits (in landed value, jobs, incomes, and other social benefits) against potential future benefits and declines in stocks (not estimable precisely) — is not a scientific or analytical question, but calls for personal or political judgment. Resolution of this balancing act is illuminated by better scientific evidence, but ultimately requires a process of reconciling the perceptions and preferences of the numerous stakeholders, each with independent and often conflicting interests in the possible outcomes.

While we talk about uncertainty and risk in both the scientific and the institutional aspects of resource management, structural uncertainty poses particular challenges. There may be indeterminacy in some components of fisheries, making it impossible to aspire to any substantial understanding of their state and dynamics. In these situations, causal chains are not easily identifiable and predicting the consequences of actions is not meaningful. We face situations in which we do not know the probabilities or odds to attach to key variables and parameters and we may not know even what relationships are critical or what questions should be asked (Wynne 1992; Gell-Mann 1994; Charles 1998). In such situations, only broadly participatory collective decisions are likely to prove acceptable to the variety of participants or "stakeholders" involved.

Moreover, in a group setting, where the scientific evidence and decision analysis must be mobilized in support of a group decision, it is necessary to recognize that data will be differently interpreted and uncertainties and risks differently perceived by different individuals, as was noted above in the discussion of risk perceptions. In this case, some interactive process of reconciliation and synthesis of beliefs and perceptions was suggested. This is a process faced regularly by such fisheries institutions as the Fisheries Resource Conservation Council in Atlantic Canada. In such instances, decision analysis may perhaps be employed to make clear the implications of alternative group decisions (Maguire and Boiney 1994).

So, while it may be convenient for expository purposes to think of risk management first in terms of analytical methods of risk assessment for supporting decisions in the presence of uncertainty, in fact (and even in principle) no such easy separation is possible. There is an iterative process whereby the judgments and perspectives of multiple stakeholders concerning the estimated risks are brought into the analysis (Jasanoff 1992; Wynne 1992; Norgaard 1994). As well, the political judgments about management options are inextricably linked to those perspectives.

Thus, it is impossible to take the uncertainty in scientific estimates of the state of the stock and the system fully into account in risk-management processes without reference to the wide range of differing perceptions of risks and conflicting interests around the uncertain outcomes. Institutional provision for broad consultation around conservation decisions is therefore necessary and these should be addressed in light of the particular policy choices to be considered against the uncertain evidence. Such an approach has been called "civic science" by the International Joint Commission (IJC 1995) (see Box 4.4).

Box 4.4. What is "Civic Science"?

"If a scientific consensus cannot be achieved and a political decision must still be made, democratic tradition requires that a range of views be expressed by experts and the public so the decision to act may be based on societal values and ethics, rather than expert calculation ...

The challenge of incorporating societal values within the two realms of uncertainty, scientific understanding and policy formulation, lies at the heart of the concept of civic science. This is the public domain where facts are interpreted and societal questions emerge as part of a great iterative process systematized by government and educational institutions ... The strength of civic science emerges when there is doubt, since it derives from all of the sciences, including social sciences and ethics, the humanities and the varied experiences of members of the community ... Such concepts as the precautionary principle and reverse onus have arisen from communities recognizing responsibility for future generations when the consequences of actions are perceived as potentially catastrophic. These strong principles of civic science are beyond the scope of the scientific risk assessment process. They imply a need for broader consultation for the decision-making or risk-management phase of policy making" (IJC 1995).

Analysis of risk and uncertainty must therefore be extended in three directions. First, it is essential to push the modelling and analysis of risk assessment and management forward to a much greater understanding of the distributional consequences of alternative management decisions (both among organizations and between current as well as future generations of people) and thus deal better with "equity." Second, the structure and operation of fishery management must be modified to reduce risks, by making management more robust to uncertainties, and less sensitive to likely sources of errors (Charles 1998). Third, it is necessary to ensure that participatory processes are also applied to data acquisition and interpretation. Thus, we need greater use of more comprehensive decision support systems, as well as greater appeal to participatory decision support processes; for example, see the work of de Leon 1992, Norgaard 1994, and Jasanof 1992.

In considering the limitations of current decision-making procedures and the possibilities for more participatory approaches, the growing literature on extended processes of multistakeholder consultation, consensus-building, and shared decisions becomes relevant. This emphasizes the importance of sustained discussion to build a common appreciation of the circumstances and shared perspectives on risks (if not shared priorities on outcomes or consequences) not only among those involved in the current commercial, recreational, and aboriginal fisheries, but potential participants in emerging and developing fisheries. Such discussions will also increase mutual understanding of the competing demands for other species (expressed perhaps as requirements to protect biodiversity), and the competing uses of habitat by humans for settlement or other purposes. Recent experience in British Columbia suggests that commitment to a sustained and demanding undertaking might be necessary to make such processes fruitful in the management of fisheries systems. One example of interest in this respect is the work of the Clayoquot Sound Scientific Panel, which operated on the basis of consensus in extended discussions which included full participation of Nuu-Chah-Nulth native elders as sources of traditional ecological knowledge.

Conclusions

We have shown in Chapters 3 and 4 that maintenance of productive fisheries and the related social and economic benefits require extensive data; however, uncertainties pervade all components of fisheries. Uncertainties arise from several sources, ranging from those in the natural and human environment, to the inherent variability and complexity of fishery systems, imperfect information about their structure, errors in data that we gather about them, and errors in implementation of policy and incomplete compliance with regulations. These uncertainties create risks to fish populations and for people who depend on them. Thus, decisions by managers, processors, fishermen, and members of coastal fishing-dependent communities could lead to reduced productivity of fish populations, reduced economic and social benefits, or both. This is particularly true because human interventions in many Canadian fisheries have attained a scale where irreversible consequences can result.

Therefore, all participants in fisheries need guidelines or principles to follow in order to avoid the detrimental effects of uncertainties and risks. The next chapter provides such principles.

Chapter 5. Principles for Sustainable Fishery Systems

Participants in Canadian fisheries face the daunting task of coping with extremely complex and changing systems about which we have only partial knowledge — leading us to make management decisions fraught with risk and uncertainty. One central message of this report is the necessity of a precautionary approach for dealing with these conditions. A second key message is that we ignore at our peril the interconnectedness of the systems we manage; we must orient our management decisions toward an ecosystem approach. Finally, our management objectives must include not only the central and overriding objective of maintaining ecosystem integrity, but also the secondary objectives of social and economic well-being. To achieve these objectives, we outline six principles that should serve to guide decision making for participants in fisheries. These principles are, in a sense, "rules of conduct" for the many activities in fisheries, from harvesting and processing, to management and policy development, and they greatly extend the principles of Olver et al. (1995).

Principles for sustainable fishery systems

The six principles are as follows. (1) Incorporate into decision-making an analysis of structural and dynamic complexities of fisheries systems. (2) Incorporate into decision making an analysis of change, uncertainty, and risk in all fishery activities. (3) Promote and conserve biological, economic, and social diversity. (4) Collect, analyze, and openly communicate data and information. (5) Estimate, document, and incorporate into decision-making the social and ecological consequences of decisions and actions. (6) Clearly define roles, rights, and responsibilities of all fishery participants to align their interests with overall objectives of sustainability.

These principles are arranged starting with those focused on inherent aspects of fishery systems and ending with those concerned with the human decision-making process. The first two principles deal with the most fundamental inherent characteristics of fisheries — their complexity and their uncertainty — and the need for all aspects of fishery operations, management, and policy to take these realities into account. The next two principles, on enhancing diversity and ensuring adequate information flow, deal with crucial tools to cope with complexity, uncertainty, and risk. Finally, as prerequisites for decision making by fishermen, industry, and managers in complex and uncertain fisheries, the full social and ecological costs of decisions must be considered in advance and institutions and management arrangements must ensure that the roles played by all participants are clearly defined, along with their respective rights and responsibilities, to align private interests with overall objectives.

For each of these principles, a full description is given below, together with an explanation of the importance of the principle and examples of what has gone wrong in fisheries when the principle was not followed. In Chapter 6, a set of strategies for implementing sustainable fisheries will be presented, with examples of experiences in applying the principles.

Principle 1. Incorporate into decision-making an analysis of structural and dynamic complexities of fisheries systems

Canadian fisheries and fisheries in general are undeniably complex. Harvests are taken from age-structured stocks of multiple interacting species, living within ecological systems, subject to natural environmental variability. A spectrum of fishermen are involved, including full-timers and part-timers, and various types of fishing gear (e.g., hook-and-line, gillnets, trawlers), both small-scale (usually inshore) and large-scale (industrial, typically offshore). Beyond the harvesting sector, fishery systems also involve processors, shoreworkers, distributors, marketing channels, consumers, government regulators, and support structures. There are more than 1200 fishing communities in Canada, in which socio-cultural factors (such as occupational structure of the communities) interact with differing economic and ecological opportunities. Fishery management agencies attempt to influence the direction of these systems in order to achieve a complex balancing of multiple objectives.

Principle 1. *Incorporate into decision-making an analysis of structural and dynamic complexities of fisheries systems.* Given their complex nature, fisheries must be understood and managed as systems. If any relevant factor in fishery systems is ignored, important interconnections may be missed, resulting in undesirable outcomes. To achieve sustainable fisheries, all participants in them must take into account a wide range of interactions: among the multiple objectives, among the ecological, economic, and social components of the fisheries, among species in the ecosystem (e.g., predator–prey effects, food chain considerations), among fishermen and fish stocks (viewing fishermen as having specific impacts on the ecosystem), among fishermen and managers (in terms of the responses fishermen make to fishery regulations), and among fisheries and their socio-economic and natural environment.

Justification

Simple recognition of complexity, while a crucial first step, is not adequate to properly respond to the structural and dynamic character of complexity. It is necessary to *incorporate analysis of structural and dynamic complexity* into the decision making process, at all levels of fishery systems. Chapters 2 and 3 discussed the range of structural and dynamic complexities of fishery systems. The implications of these complexities are presented in Chapter 4. Although modern theory on "chaos" and "complexity" suggests that it may be impossible to fully understand and predict the behavior of fisheries, there is nevertheless a clear need for analysis of complexity and for accounting for it in fishery decision making. The interdisciplinary approach taken here illustrates the need for an integrated approach to research and information acquisition, to better take into account the structure and dynamics of the system and the many interactions among its components. Given our lack of control and understanding, fishery management should be designed for robustness, to be reasonably successful even in the presence of complexity.

We collect very little information about fishery systems, especially in comparison to the scale of the problem. In assessing fish biomass, for example, we survey only a small percentage of the fish that we directly harvest. In sampling the physical ocean, much attention is paid to the continental shelves of the northern hemisphere but even today, after a century of effort, many parts of the ocean are rarely sampled. We must therefore keep in mind that we normally undersample fisheries. Furthermore, the economic and social systems receive scant attention and the attention is not coordinated when they do. Thus, the problem is both complex and enormous. It is unlikely that we will ever have adequate information to provide a detailed description of an entire fisheries problem. Nonetheless knowledge of this complexity is important.

Accounting for these considerations is necessary, given this undersampling, and may be as simple as a qualitative acceptance of our ignorance of parts of fishery systems or constraining our enthusiasm for overinterpreting or predicting dynamics of fisheries based upon limited information. Since the goal of complete knowledge is unobtainable, we must at least explicitly account for its incomplete character.

What happens if we do not take complexities into account?

A failure to take complexities into account in fisheries has manifested itself in many ways, ranging from problems arising from single-objective thinking, neglect of multi-species, multi-stock dynamics, failure to appreciate the role played by coastal communities, and the links between fisheries and the coastal zone. A selection of these various themes is addressed here.

First, consider the complex mix of objectives pursued by the various players in fisheries. These goals range from "economic efficiency" to social and community stability. If the overall benefits to society are maximized through policies that balance a variety of socio-economic objectives, then the best policies will be ones that achieve such a balance while also ensuring ecological sustainability. However, it is common for fishery policies to focus excessively on just one objective (such as short-term economic efficiency), which tends to produce fishery policy lacking a suitable balance. Furthermore, if substantial numbers of stakeholders have objectives different from those pursued by managers, decision makers may well fail to achieve anyone's objective adequately. Thus, it is crucial for managers to understand, not merely to assume, the objectives being pursued by fishermen. For example, the frequent but simplistic assumption that fishermen are profit maximizers is not necessarily true. It is common for decisions by individual fishermen to be made not with profit-maximizing formulas, but rather using *ad hoc* rules of thumb; for example, investment decisions may be based on such rules as "one quarter of gross revenues go to the boat."

Second, immense advances in trawler technology in the Atlantic Canadian groundfish fishery have had a major impact on that fishery, in part because these were not taken fully into account in fishery management. While attention focused on monitoring one part of the system, biomass levels and catches, too little effort was placed on understanding and monitoring the complex behavior of the fish harvesting activity itself, in particular, how fishing efficiency was increasing over time. There also are interactions among these components. The setting of quotas requires a knowledge of the fish biomass, one index of which (in some stocks) is the "catch rate," the rate at which fleets catch fish (per unit of fishing effort). By failing to understand and monitor how effective effort was changing over time, use of the latter catch rates for Atlantic groundfish proved misleading. Partly through technological improvements in catching efficiency, fishermen were able to maintain catch rates even as stocks declined. High catch rates gave the illusion of strong stocks, so that excessive quotas were set and overfishing resulted.

Third, a focus on single-species management has meant that complex interactions between species have received little attention. For example, even though capelin, herring, and other sources of food for groundfish are caught commercially, there is little knowledge of the impact of such harvests on the abundance of depleted groundfish stocks. Furthermore, within a species-specific approach, relatively little attention has been paid to the quality of fish habitat and in particular the impacts of fishing activity on fish habitat. Many fishermen believe that such failures to account for ecosystem complexities contributed substantially to groundfish collapses, although the lack of research attention in these areas means that such concerns are difficult to test.

Fourth, a failure to consider stock structure has created problems for fisheries on both Canada's Pacific and Atlantic coasts. The multistock nature of Pacific salmon stocks is well-known, but excessive attention to the most productive of these stocks has led, over the years, to a loss of genetic diversity as less productive stocks disappeared, almost unnoticed. On the Atlantic coast, the northern cod stock was typically treated as a single entity, despite evidence that it may have included several very different substocks. The evidence for stock structure is still difficult to obtain, particularly for migrating stocks that cover large geographic areas, but some fisheries managers have been overly eager to apply simple models of stocks rather than work towards a more ecologically realistic management model. Such models may be difficult, indeed they may be beyond our present grasp, but we should still advance as far as possible the analysis of multiple stocks.

Fifth, Canadian fishery management over the past 25 years has been based on administrative units of management that, while dealing daily with the complexities of gear types, fleets and harvests, do not sufficiently consider the nature of fishing communities. As a result, the statistics produced from these administrative units hide a good part of the social diversity of fisheries systems. In particular, as a practical consequence, more emphasis tends to be placed on capital in fisheries ("fleet management") than on the differing strategies and behaviors of human beings (the people behind "labor"). There has also been a lack of understanding of how community institutions can assist in improving the effectiveness of fisheries management. Overall, the complex structure of fisheries systems, and notably of fishing communities, is not taken into account in addressing matters of access, management, monitoring, and research.

Finally, many coastal fisheries, in Canada and globally, face a trio of fundamental problems: overexploited resources, overextended fleets, and a lack of employment options outside of fishing-related jobs. Unfortunately, fisheries policy tends to say little about the latter problem, having been developed and implemented without understanding the dynamics of coastal economies. Just as any given fish population interacts with others in the ecosystem, so too does fishing interact with other components of coastal economies. As analyzed in Chapter 2, efforts to deal with excess capacity in fisheries that try to remove "surplus" fishermen typically fail because they do not deal with the lack of alternative local employment opportunities; there is no alternative employment for people displaced from these fisheries and displaced effort is typically redirected at other fisheries.

Nevertheless, many current problems of fisheries management in Canada require an appreciation of the complexities inherent in fishery systems and as such can be approached only through appropriate interdisciplinary research that includes anthropology/sociology and political science, as well as the more traditional topics of economics, and biological and physical sciences. Understanding of the relations among fishermen, scientists, and managers is crucial for improving management options and represents a key issue for the future of fisheries. Recent fishery collapses have revealed that the specialized analyses of the past do not provide the range of concrete solutions necessary to reestablish or maintain an equilibrium between preservation of the environment and the needs of the harvesters.

Principle 2. Incorporate into decision-making an analysis of change, uncertainty, and risk in all fishery activities

As noted in previous chapters, there is great uncertainty in components of fishery systems: fish stock status, ecosystem structure and function, human behavior, and the potential effects of alternative management strategies on fisheries resources. These uncertainties create risks — ecological risks (e.g., loss of stocks), economic risks (e.g., loss in economic yield from harvests), and social risks (e.g., social dislocation in fishing-dependent coastal communities).

> ***Principle 2. Incorporate into decision making an analysis of change, uncertainty, and risk in all fishery activities.*** To be effective, analyses, planning, decisions, and actions by everyone involved in fisheries must recognize and take into account the diverse sources of uncertainty and their resulting risks. This is true for decision making by fishermen and others in the fishing industry, as well as by fishing-dependent communities and by fishery management agencies.

Justification

The basic rationale for taking uncertainties into account is that a failure to do so is likely to lead (and has led in the past) to inappropriate management institutions or improper decision making. In particular, the risks noted above are often not properly or commonly considered by decision makers. Here we identify two principal means for taking uncertainty in fishery systems into account: adoption of a precautionary approach and appropriate design of institutions and frameworks for participatory decision making. We first discuss the precautionary approach, which emphasizes fisheries operations and management that reflect the variability, uncertainties, and risks discussed in Chapters 3 and 4.

The concept of a precautionary approach to marine resources has been under development for many years, especially in Europe, and has recently been formalized and expanded upon by the Food and Agriculture Organization of the United Nations (FAO 1995*c*, 1995*d*). "The precautionary approach involves the application of prudent foresight. Taking account of the uncertainties in fisheries systems and the need to take action with incomplete knowledge, it requires, [among other things] a consideration of the needs of future generations and avoidance of changes that are not potentially reversible" (FAO 1995*c*, p. 4). We emphasize six characteristics of a precautionary approach as follows. (1) Unavoidable uncertainties (from natural variability and measurement error) should be recognized and considered explicitly when management agencies seek appropriate regulatory policies and when members of the fishing industry make investment decisions. (2) Management agencies and the fishing industry should act to reduce the magnitude of uncertainties. (3) To maintain benefits from fishing over the long term, "... where the likely impact of resource use is uncertain, priority should be given [by management agencies and industry] to conserving the productive capacity of the resource" (FAO 1995*c*, p. 4). (4) Steps should be taken to preserve and maintain the health of critical habitat. (5) The precautionary approach also requires that "harvesting and processing capacity should be commensurate with estimated sustainable levels of the resource, and that increases in capacity should be further constrained when resource productivity is highly uncertain" (FAO 1995*c*, p. 4). (6) Undesirable outcomes of management actions should be identified prior to the onset of fishing, and agencies must ensure that ... "any necessary corrective [regulatory] measures are initiated without delay and that they should achieve their purpose promptly" (FAO 1995*c*, p. 4).

Other features of the precautionary approach that have been recognized internationally are included elsewhere in this chapter under Principle 3 (enhancing biological, social, and economic diversity), and 4 (ensuring adequate collection, analysis, and distribution of data).

It is important to differentiate between the precautionary approach and the precautionary principle. In fisheries-related discussions (e.g., Garcia and Newton 1997), the precautionary principle refers to a situation where an extremely restrictive action is taken, such as completely shutting down a fishery, as a safeguard measure, based on an assumption that the fish stock cannot withstand any disturbance — even when that assumption is based on weak evidence. In contrast, the precautionary approach involves some estimation of potential impacts of each harvest rate before determining which rate is appropriate. Many different rates (only one of which would shut down a fishery) would be considered possible and their outcomes would depend on the magnitude and nature of the uncertainties, among other things (FAO 1995c. In extreme circumstances (such as occurred for Atlantic Canadian groundfish in 1993), the need for rapid action may call for application of measures referred to under the precautionary principle but the discussion here focuses on the precautionary approach. [Some other documents, such as Agenda 21, use "precautionary principle" in the more general sense referred to here (and in much of the rest of the literature) as the "precautionary approach."]

The other important element of Principle 2 is to properly design the framework for interactions among all "players" in fisheries. For instance, in order for fishermen and processing companies to appropriately take into account uncertainties and risks associated with environmental variation, stock dynamics, international markets, and other such factors, they need full information on the uncertainties and risks. In particular, they need to understand in detail the *long-term*, as well as short-term, implications of their various actions, for example, discarding fish, purchasing electronic equipment to increase their search efficiency, or investing in new processing plants. Catching more fish may seem wise in the short term for fishermen, but in the long term, the negative consequences could appear if a stock is overharvested. Likewise, managers who must make tradeoffs when choosing among alternative regulations must also clearly understand the implications of their choices.

In all instances, clear communication of the inherent risks is essential to permit people to make appropriate decisions. As suggested in Chapter 4, the technical capability to estimate risks usually resides mostly in stock assessment units of fishery management agencies. But even within these institutions, there is often a communication gap, such that the information emerging from risk assessments is frequently not clearly understood by those who need to use that information. This means that considerable effort should be expended by risk assessors to ensure that information on risks is produced in forms that are understandable even to non-technical experts (Covello et al. 1986; Morgan et al. 1992; Rowan et al. 1994). There is excellent research emerging from cognitive psychologists about dealing with such communication problems.

Finally, knowledge about risks (economic, social, as well as biological) is also obtainable from fishermen and others with local knowledge. This information needs to be integrated into analyses by everyone, including government decision makers. Through an iterative participatory process of analysis and evaluation of options, all participants in fisheries should be able to make more appropriate decisions with more complete understanding of the consequences of their actions in terms of the risks. This will help to achieve the overall objectives of having productive aquatic ecosystems, economically viable fisheries through sustainable harvests, and socially stable fishing-dependent human communities. Some of these ideas about information are elaborated further under Principle 4 in the broader context of the general need for better collection, analysis, and communication of information in fishery systems.

There are important interactions between the precautionary approach and better design of institutions and communication mechanisms. For instance, change, variability, and uncertainty exacerbate the problems of overcapacity and a short-term focus, in part by obscuring the risks of harvesting or other human activities. This contributes to the lack of understanding of the risks, or expected losses, that may result in the long run from inappropriate decisions. However, an underlying tenet of the precautionary approach deals with this lack of consideration of long-term risks in favor of short-term gains by emphasizing prudent foresight, consideration of the needs of future generations as well as the present one, and avoiding changes that are potentially irreversible. This can be done by explicit consideration of forecasts of long-term risks before decisions are made by management agencies, the fishing industry, or fishing-dependent communities. For example, management agencies should set harvest rates by explicitly considering relevant uncertainties and the resulting risks to stocks. This will bias actions toward minimizing risks and maintaining populations in a productive state over the long term. As well, industry and communities should realistically evaluate proposed capital investments in the context of the known variability in fish stocks and their habitat and the uncertainty in estimates of status of stocks. If these views are used as guidelines by all people involved in fishery systems, this will improve prospects for long-term sustainable fisheries.

Harvesting of many long-lived fishes needs to be especially precautionary (e.g., less size-selective and at a reduced harvest rate) because in such cases the age structure can be changed radically from the unfished state so that there are fewer large, old fish, as described in Chapter 3. Such large fish produce a disproportionately large number of offspring per parent (Trippel 1995) and the scarcity of such spawners both reduces the reproductive output of the stock and increases the chance that a population will not be able to persist through a lengthy period of detrimental environmental disturbance. Because such environmental variation is common (Chapter 3), harvesting schemes should recognize the need to preserve the productive "capital" in the stock by imposing size limits and stopping fishing in spawning or nursery areas. Furthermore, for stocks that are just recovering from depletion, if priority is not placed on preserving strong year-classes for future reproduction (rather than current harvest), large future economic and social benefits could be foregone.

What happens if we do not take into account change, uncertainty, and risk?

The need for recognizing and taking into account change, uncertainty, and risk has become increasingly evident in the wake of significant declines in several important global fishery resources (FAO 1995*a*; Gordon and Munro 1996). Among the factors contributing to widespread overexploitation of marine fishery resources has been the tendency by fishermen, industry, and government managers to adopt risk-prone decisions in the face of uncertainty, for reasons noted above. Application of the precautionary approach represents an important corrective force, reversing the traditional burden of proof so that it is placed not on scientists and managers but on users of the resource. A common outcome of applying a precautionary approach to fishery management is that the greater the uncertainty, the greater the need for caution. The legacy of decimated fishery resources on a global basis points unequivocally to the disastrous implications of ignoring uncertainty and the consequent need to adopt a precautionary approach.

Unless a precautionary approach is taken, which includes taking uncertainties explicitly into account, the resulting choice of harvest levels may increase the probability of serious and perhaps irreversible depletion of stocks. As noted in Chapter 3, biological as well as physical environmental variation can cause fish growth and survival rates to change rapidly and in some cases by large amounts. Similarly, technological advances usually lead to increasing efficiency of each boat-day of fishing effort. There has been too little effort on the part of management agencies to monitor and take into account such changes and more generally the changes resulting from human responses to fishing regulations. Thus, we have a situation in which uncertainty is increased, since imperfect information prevents management agencies from reliably detecting such changes. Furthermore, there is imperfect control over the human component of fisheries systems. Thus, these factors of change and uncertainty combine to create substantial uncertainty about appropriate harvest rates and this needs to be taken into account when setting regulations.

There are many cases where these and other undesirable outcomes have occurred, at least in part because a precautionary approach was not taken. For instance, the fish stock that once generated the world's largest annual tonnage of catch, the Peruvian anchovetta, collapsed in 1972 due to a combination of several factors: (1) failure to restrict large capital investment in fishing and processing capacity, (2) failure to keep young (even prereproductive) fish from being harvested at high rates, and (3) failure to close the fishery early when environmental conditions became unfavorable for the fish (an ENSO event — El Nino – Southern Oscillation). The northern cod population in eastern Canada also diminished rapidly after several years in which harvest rates were higher than the target, when the age structure became narrower, and when environmental factors became unfavorable (Hutchings and Myers 1994). Murphy (1977) and Saetersdal (1980) also review case histories of collapses of several important herring, sardine, and anchovy fisheries in Europe, Asia, Africa, and North America. In many cases, insufficient action was taken to restrict disturbance and harvesting of spawning populations or to prevent harvesting of fish prior to their first age at maturity. As well, many of those collapses were accompanied by environmental change. Although there are different hypotheses about the ultimate cause of the reductions in abundance (e.g., environment vs. fishing), both mechanisms contributed to some degree (Murphy 1977); without the effect of fishing, some of the stocks may not have collapsed when they did or at all, and similarly, without unfavorable environmental conditions, other stocks may not have collapsed when they did or at all. These and other well-known examples illustrate the potential risks associated with not taking precautionary measures.

Finally, stocks and human livelihoods are at risk unless appropriate measures are taken to recognize uncertainty and risks involved with excess harvesting capacity. There is now little margin for error in manipulating harvest rates (Rosenberg et al. 1993). Furthermore, chronically undersatisfied fleets and processors may have aggravated a tendency towards excessively liberal total allowable catches and exacerbated the problems of monitoring, control, and surveillance. Characteristic 5 of the precautionary approach described above addresses this overcapacity issue by requiring that "harvesting and processing capacity should be commensurate with estimated sustainable levels of the resource ..." This will reduce pressure by industry on managers to harvest at levels above what they should be, which is particularly important given natural variability and lack of complete understanding of the fishery in question.

Of course, reduction in harvest capacity has proven very difficult in practice. Thus, pending the achievement of this elusive goal, it is crucial that harvest rate (the realization of that capacity) be properly controlled. A reliance on quota (harvest quantity) controls by government agencies, as has been the case in many fisheries, is insufficient to ensure sustainability, since it increases the probability of accidental overexploitation and encourages anticonservationist practices, such as dumping and discarding (Copes 1997; Charles 1995, 1998).

Principle 3. Promote and conserve biological, economic, and social diversity

At the simplest level, biological diversity (biodiversity) is defined as the degree of variation in the numbers and kinds of species and ecosystems within or among regions at a given time (Savard et al. 1994). More precisely, biodiversity has been expressed as the collection of genomes, species, and ecosystems occurring in a geographically defined region (NRC 1995 cited in Boehlert 1996) or "the sum and interaction of the variation that exists among populations, species, communities, and ecosystems" (Philipp et al. 1995). In an analogous manner, human (social and economic) diversity can be viewed as the degree of variation, or "richness," in economic and social entities within a region. This might be reflected in the existence of diverse livelihood/employment opportunities, diverse coastal communities (each with its own special characteristics), or diverse management approaches for conserving and exploiting natural resources. (Note that here, human diversity is viewed from a fishery perspective, reflecting a variety of potentially conflicting as well as common interests, rather than in terms of genetic differences among humans.)

> ***Principle 3.*** *Promote and conserve biological, economic, and social diversity.* Conserving and (or) enhancing biological, economic, and social diversity within fishery systems provides an important hedge against uncertainty. Measures that maintain biodiversity are also inherently important to maintaining long-term economic viability and flexibility, as well as social adaptability and stability. It is therefore important to adopt options, policies, and incentives that are compatible with maintaining biological diversity (at the levels of fish populations, species, and ecological communities) and socio-economic diversity (at the levels of human communities, and local and regional economies).

Justification

In the pursuit of sustainable fisheries, it is crucial to maintain both biological integrity (reflected by the actual dynamic functioning of the biological system) and biodiversity (Winter and Hughes 1997). We focus here on the second of these (which in fact is simply an indirect index of the functioning of biological systems, since it reflects a structural attribute, rather than a functional one generating their dynamics).

Research on ecological systems has shown that although there is not a clear, uniformly applicable increase in stability with increasing diversity, certain ecological systems do show increased persistence and integrity of functioning with increasing diversity. However, well-controlled experiments are not possible with real systems, so ecologists err on the side of caution when suggesting that biological diversity be maintained or at least not allowed to deteriorate severely. Natural systems, including marine systems, have been subjected to natural fluctuations for millennia and the systems that we now harvest have, by definition, evolved mechanisms to adapt to those variations. Their resilience to disturbances is in part due to the structural and dynamic complexity described previously. However, a precondition for structural complexity is that the systems be diverse at several levels: genetic, species, and community level. That is, a population with a wide range of genetic information among individuals is more likely to adapt successfully to changing environmental conditions than is a population with greatly reduced genetic diversity due to strong selection. Witness the numerous examples of single-variety agricultural crops suffering from massive outbreaks of diseases. Similarly, a community composed of many species, some of which can fulfill the functional role of other species, is more likely to persist through periods of considerable environmental change.

Because we do not know the adaptive value of particular genetic or species attributes, actions that lead to decreased diversity within populations or among species may lead to their decreased resilience when faced with ecological change. The populations and species represent the only reserves of genetic variation that have allowed their persistence and thus there is no real alternative to conserving that diversity (Ryman 1991). The value of conservation of diversity lies in its ability to preserve future options not only for species, but also for those who exploit them.

These ideas are the basis for the current widespread concern for losses in biodiversity, which are being driven by the continuing increase in the size of the human population and the per capita consumption of resources — both occurring at rates that preclude long-term resource sustainability. It is feared that biological systems will lose their ability to adapt to variations caused either by natural or human induced changes and that those systems will therefore not persist in their current productive state. In the case of marine ecosystems, this could mean that fisheries and the economic and social benefits of them could be jeopardized by significant losses in biodiversity.

Discussion of this principle focuses separately on biodiversity (particularly genetic diversity) and on human (social and economic) diversity.

Biodiversity

The most important ways by which genetic diversity within a species or among its stock or population components is threatened include (Ryman et al. 1995; Philipp et al 1995): (1) local population extinction that results from excess exploitation, habitat loss, disease introduction, and species displacement; and (2) intentional or unintentional selection pressure coincident with a reduction in population size. Thus, overfishing, habitat degradation, and hatchery programmes, especially in conjunction with one another, lead to the loss of genetic diversity (Winter and Hughes 1997).

At the community level, selection directly or indirectly imposed by fisheries on assemblages of species are very similar to those that act on the life-history characteristics of specific populations (Chapter 3) but operate at the species level, as opposed to the individual and population level. In general, selective fishing will favor species that have certain characteristics such as small maximum or terminal size, early size and (or) age of maturity, high fecundity, and high growth rate. Such characteristics generally describe the smaller pelagic species. For example, intense fishing pressure on Georges Bank has resulted in a relative shift from the dominance by groundfish species.

What happens if we do not conserve and enhance biological diversity?

Thorpe et al. (1995) have summarized the genetic processes associated with human activities related to wild fish harvesting and aquaculture, many of which create high to extreme risk. Clearly, environmental change in concert with selective harvesting and (or) overharvesting adds to that risk.

Modern fishing practices have a direct effect on species composition and interactions among species by the removal of target and non-target (through by-catch) fish as well as through the physical disturbance of habitat (reviewed in Boehlert 1996). There are additional indirect and complex interactions that alter predator–prey and trophic relationships. Thus, severe depletion of marine species or stocks may change the function of the ecosystem and the flux of energy and materials within it. Furthermore, fishing-related and other stressors (eutrophication, species introductions, habitat degradation, and contamination) tend to drive aquatic ecosystems to simpler, less resilient states (Ryder and Scott 1994). Thus, overall ecosystem productivity, energetics, nutrient cycling, and community structure may be altered (Upton 1992; Ryder and Scott 1994; Boehlert 1996). Loss of habitat implies the loss of the part of the genetic diversity of a stock that would have used that habitat (Thorpe et al. 1995). It also follows that the more limited the species diversity constituting an ecosystem or region, the more limited that ecosystem will be in sustaining itself in the face of change. Thus, as biodiversity is reduced, our current and future socio-economic options (short- and long-term) become more limited.

Social and economic diversity

The conduct of Canada's fisheries and the well-being of its 1200 fishing communities rely on maintaining both biodiversity and socio-economic diversity; these lie at the heart of fishery dynamics and development. Just as biodiversity reflects the degree of variation in the numbers and kinds of species and ecosystems, socio-economic diversity is a measure of similar variation on the human side of fishery systems. This includes economic diversity (such as the extent of types of fisheries and employment alternatives) and social diversity (e.g., the range of coastal communities in a region), as well as diversity in fishery objectives and in management approaches.

Emphasis here is placed on economic and social diversity, but we first provide a comment on diversity in fishery objectives and in management approaches. Recognizing diversity in fishery objectives prevents an excessive focus on single-objective policies, such as complete privatization or short-term profit maximization, which may be imposed at the expense of other more resilient strategies (e.g., note the widespread expansion of aquaculture, despite its increased ecological and social costs.) The need for diversity in management approaches is clear in that it provides robustness against the possible failure of any one approach and allows appropriate adaptability of management to specific regional or local requirements.

What happens if we do not conserve and enhance economic and social diversity?

What are the consequences of low economic diversity (a lack of employment opportunities)? If this occurs in conjunction with serious depletion of the relevant fish stocks, so that fishermen are unable to continue harvesting, the economic and social consequences frequently lead to increased government responsibility and intervention, or, where government does not intervene, to socio-economic instability for fishermen and their communities. On the other hand, if fishermen are still able to obtain a profit from fisheries, they tend, due to limited options, to keep their effort at the same level even though they are uncertain of the future conditions of the resource they exploit. Thus they face a significantly increased risk to investment and income in the medium and long term. If this situation prevails where licensing and quota programs have increased inequality among various groups of fishermen, the difficulty of establishing consensus on necessary management measures will be highly destabilizing.

What are the consequences of failing to recognize social and economic diversity in fisheries? A failure to incorporate the socio-cultural importance of fishing to local communities and to involve those communities in management results in a dysfunctional working relationship between the community and others involved in fishery management, be they scientists, administrators, or politicians. There is a lack of positive incentives.

Energy is put into conflict resolution rather than into rational planning of fishery systems.

A loss of economic diversity can arise from the drastic decrease of small-scale fishing units operating from rural coastal communities, a trend on the west coast. Research indicates that since rural fishermen are more likely to live close to the resources (and habitats) they exploit, they are likely to have a relatively greater knowledge of and identification with the resource, as well as greater incentives to invest in monitoring, restoration, and enhancement. Unfortunately, some communities with old fishing traditions or rights, such as aboriginal communities on the upper Fraser River, have been deprived of an allocation of fish for so long that a generation has grown up with no experience or (in some cases) interest in fishing. The latter group has been alienated from fishing while many urban fishermen may have become alienated from the basic productive conditions (the habitat). If such well-placed "guardians of the habitat" are not given roles in management and rights to protect habitat, they will have little incentive not to use the habitat for other income-generating purposes. A loss of economic diversity can thus contribute to the loss of fish habitat.

A loss of social diversity (through a decline in the welfare of key coastal communities) can have major impacts on biological diversity. For example, in British Columbia salmon fisheries, the greatest economic benefits accrue from a few major stocks and fisheries of the Fraser and Skeena rivers, which in turn capture the majority of management attention. Most other stocks are managed passively and in some cases not monitored at all. As a result, nearly half of the more than 9000 salmon stocks in B.C. and the Yukon are of unknown status (Slaney et al. 1996). It is virtually certain that many of these stocks and the biodiversity they represent will be lost without greater attention to monitoring their status (DFO 1988). As government agencies cannot cope with such monitoring, there are significant opportunities for involvement of communities that are strategically located to exercise stewardship, to monitor the stocks, their habitat, and related area fisheries. Thus the continuation of an economically viable capture fishery may itself be dependent on maintaining social diversity, by keeping coastal communities in many isolated regions of B.C. economically viable. (It is through such interest in capture fisheries that many community-based salmon enhancement programs have been able to secure volunteer labor from some 8000 volunteers in B.C. communities.)

The loss of the social diversity of coastal communities (if these were to be eliminated) may possibly affect the future diversity of wild salmon stocks because of the role some communities now play in attempting to conserve smaller local stocks by opposing the placement of salmon farms in their local areas in B.C. The major processors are proposing large-scale expansion of their salmon farming operations. Their "two rivers strategy" (production of major wild sockeye salmon on the two large rivers, the Fraser and the Skeena, and fish farms everywhere else) is opposed by the fishing-dependent communities in isolated parts of the coast, on the grounds that salmon farms pose unacceptable risk to wild stocks (through spread of disease and loss of habitat). Besides environmental groups, coastal communities are the main constitutency opposing this expansion, through their wish to protect the diverse wild stocks in hundreds of small systems in local areas. Of course, coastal communities at risk are not completely immune to promises of wealth held out by salmon farmers, but they are also aware of the high level of risk and the irreversability of losses of wild stocks.

At the same time, a loss of diversity in fish harvests can affect economic and social diversity. In particular, the cod and other groundfish moratoria on the Atlantic coast and restrictions on salmon fishing on the Pacific coast illustrate well the severe consequences for coastal communities of failing to protect biodiversity. For example, currently in Atlantic

Canada, traditional groundfish species have lost their past importance in many areas, where economic well-being is more and more dependent on a small number of species, notably crustaceans such as shrimp, crab, and lobster. Such restricted alternatives in fishery production make it all the more important for all interested parties in fishery systems to work more closely in establishing policies that use diversity to their advantage.

In the past, government has made efforts to explicitly protect social diversity, through development and enhancement of rural coastal communities. For example, fishing licenses were given out earlier in the century as a means to encourage settlement of the northern B.C. coast. Over a period of fifty years, federal policy and funding facilitated the entry of isolated First Nations communities into the commercial fishery. Rural development has been supported by other federal policies as a hedge against economic shocks, and to support Canadians who prefer rural lifestyles, and (or) who are better off both economically and socially in small towns or rural regions (Sinclair and Felt 1993). Indeed, a diversity of approaches to fishery management (and habitat restoration, protection, and enhancement) also supports social diversity.

However, it appears that fleet rationalization policies have operated contrary to other policy goals, often having the opposite effect. For example, on the Pacific coast, the result is a continuing shrinkage of the number of small gear licenses in the salmon fleet which has a disproportionate effect on rural coastal communities. Most recently, formerly diversified small gear rural fleets are shrinking in most areas because they were reclassified as single-gear that have to purchase other licenses just to maintain their former diversity. In other words, fleet rationalization policies do not necessarily reward small-scale diversified efficiency but reward access to capital. Those who have made large profits in any fishery or other operation have a tax incentive to reinvest in licenses. Meanwhile, on the Atlantic coast, beneficial policies of "multi-species licensing," to encourage diversification of fishermen's activity to a variety of species, have not been consistently promoted by government. Furthermore, diversification by fishermen into non-fishery employment has been actively discouraged by government, through measures that discriminate against part-time fishermen.

Principle 4. Collect, analyze, and openly communicate data and information

Principle 4. *Collect, analyze, and openly communicate data and information.* Proper management and planning require timely and adequate data. This is particularly the case within a changing and uncertain natural and human environment. If the best decisions are to be made by fisheries management agencies and other co-managers, data are required for all components of the system: the fish stocks, the physical and biological parts of the ecosystem, the fishermen, the fishing communities, the fishing companies and markets, and the global economy. The collection of adequate data requires appropriate institutional support for such efforts but also appropriate coordinated planning in the development of experimental designs for observational programs. Furthermore, in the past, too little emphasis has been given to the full richness of the required data. Until now, much of the data collected has been guarded jealously, with too little open communication; free and open flow of information is crucial.

Justification

Obtaining adequate information is clearly the first step in the decision making process. There are several steps in the flow of information: design of an observation program, acquisition of the data, analysis, and communication. Sometimes simple observations are easily interpreted on their own, e.g., the presence or absence of fish, but at other times sophisticated analysis may be required to make use of raw observations. To date the DFO has been the primary agency responsible for the collection and interpretation of most fisheries data. As a result they have also held primary responsibility for determining what data to collect. Under the Oceans Act and the revised Fisheries Act, DFO appears likely to maintain this primary mandate. As argued throughout this report, there is, however, a clear need to broaden our view of the fisheries to move beyond the fish and the environment and to encompass the social and economic components of the fisheries. As well, more players need to be included as participants.

Complete information is not possible for a system as complex and diverse as fishery systems. Thus even with much greater financial support for its monitoring and study, it is unreasonable to expect perfect information. Relative to the scale of the problem, we collect only a tiny amount of data. Our interpretive analysis must adapt to this constraint; however, the constraint points to the pressing need to maximize the efficiency and efficacy of our efforts. The situation is exacerbated by the present limitations on funding and the likelihood that funding is unlikely to increase dramatically in the foreseeable future. Indeed, over the past several decades we have lost a number of monitoring programs, e.g., the oceanographic sampling at the weather stations off the east and west coasts. The few remaining long-time series in place are often overused because they provide such powerful information about the long period changes of the fisheries. Their importance highlights the need for coherent sampling over many years versus the more common sporadic and uneven sampling of fisheries systems. Such losses highlight the precarious position in which we presently find ourselves, with a shrinking pool of available information.

In light of these limitations, it is clear that we should make use of all available data. We need to move beyond the natural sciences. Thus data must be collected on the social and economic systems in addition to the fish and their environment. Collecting information on the social and economic system would help us understand the fishermen and the environment in which they work and live. We also need to consider the local knowledge which various players in the system can provide. In the past such information has been ignored, because of its anecdotal nature, making it difficult to quantify. This information needs careful, integrated collection and analysis, but then so too do all the other types of data to which we refer.

The diversity of fisheries is another reason for the need for transparency because of the biases that each actor brings with them. Thus, government agencies, community groups, fishing companies, university researchers, and non-governmental organizations all bring with them some world view that colours their use of data. Transparency and openness of communication will help to balance these different viewpoints so that collectively we gather the most appropriate data and maximize our use of it. Open communication of information also leads to the development of consensus as the data are shared and interpreted collectively. Without such a process, any incentive scheme to encourage compliance with regulations, no matter how sophisticated, is unlikely to succeed, because distrust will almost inevitably develop, as we have seen, for example, in the early 1990s following the collapse of the northern cod stock.

What happens if there is not appropriate acquisition and flow of information?

Data collection requires careful planning and thoughtful analysis. The end use of the data must be considered before data collection begins or most of the data is likely to be irrelevant. Raw data, poorly documented, and collected in a haphazard fashion without some clear hypothesis or use in mind is unlikely to ever be of any value. Environmental data collection is replete with examples of such time series, collected with the best of intentions. It has often been the case that different data collection programs were blissfully unaware of each other. Thus different teams responsible for different fish species in the same area would not share data with each other, leading to lost opportunities for the discovery of related patterns and trends (Hutchings et al. 1997) . Thus, some of the effort in the collection of data has been ineffective because of inadequate planning and coordination.

In addition to uncoordinated collection problems, the failure to share data has meant that opportunities for interpretation have been lost. Without integrated data analysis, alternative views of complicated systems may not emerge. In the absence of transparency, a characteristic of open information flow, crucial information be lost or be unavailable to a group that could provide useful input to the analysis and decision making.

Data that are held back too tightly may also lose their value in the long term because, over time, important characteristics of the data are forgotten. Both fishing companies and government agencies are guilty of hiding information under the guise of economic or political exigency. University researchers can be guilty too, holding back data that they want to use in research and for publication. At times it may be reasonable to withhold data, but there must be some timetable for its release to ensure that it is not lost and to let all those involved know that the data will eventually be available for open scrutiny.

An unwillingness to share data openly naturally leads to distrust, not only among different researchers, but also between government and fishermen. In the past, the setting of fishery quotas has been a backroom operation with little of the data released openly. Misinformation, or misperceptions, passed on to the industry can lead to serious investment errors and the creation of overcapacity. This process may be changing, but there remains residual distrust based upon this secretive approach to public information. New programs such as the sentinel fishery on the east coast, in which fishermen collect oceanographic data and fish, point to new directions in the joint collection and interpretation of fishery data.

Limits on the types of data collected are also a problem, e.g., the previously mentioned lack of social and economic data on fisheries. The current model for fishery management is to collect data on the fish and the environment and then develop management strategies to preserve the stock and maximize economic benefit. The European Social Sciences Fisheries Network has a more integrated approach and recently called for the development of a social sciences database for fisheries, similar to those that have been developed for the biological and physical data. Such a strategy could be developed and applied here in Canada but needs to be coordinated with other aspects of the information management scheme.

Principle 5: Estimate, document, and incorporate into decision-making the social and ecological consequences of decisions and actions

It is important to explicitly account for the full social and ecological costs of decisions and actions in fisheries. For example, a fishery management decision that is likely to be detrimental to fish habitat will have a long-term ecological cost. Action that affects the well-being of coastal communities implies a social cost. Both of these costs may have been missed in traditional calculations used in past decision making, but must be incorporated in the future.

Principle 5. *Estimate, document, and incorporate into decision-making the social and ecological consequences of decisions and actions.* There is a need for "full-cost pricing" that reflects the full impact of decisions, not only monetary, but social and ecological as well. This recognizes that sustainable fisheries rely on maintaining both the natural capital — the fish stocks and the environment in which they live — and the social capital built up in communities and regions.

Justification

Fish in the sea represent a public resource owned by all Canadians, present and future. The wise use of these fish can be directed toward meeting a range of current social and economic goals, while protecting the resource and the ocean ecosystem for use in the future. To determine what constitutes wise use of fishery resources requires full accounting for the social and ecological effects of present-day exploitation patterns, including both the harvest *level* and the harvesting *methods*.

For example, a harvest that is too large this year is likely to have a negative impact on the ability to achieve reasonable catches in the future; this represents a social cost, imposed on the future. Alternatively, an apparently "reasonable" harvest taken in a damaging manner, perhaps by focusing on juveniles, or by disturbing the ocean bottom, imposes an ecological cost. A management decision to favor capital-intensive fishing, for example by allocating half of the Atlantic groundfish harvest to offshore vessels, would have a social cost through detrimental impacts on hundreds of inshore fishing communities and thereby on the coastal economy. Pursuit of fleet rationalization (reduction) through license stacking or transferable quotas imposes similar social costs, through unplanned concentration of licenses among a few owners and increased unemployment.

Why have ecological and social costs not been taken properly into account in the past? There appear to be several key reasons. The first is fundamental human attitudes. Oceans were treated for years as the last frontier for unbridled exploitation and the entreprenuerial spirit. Ecological and social costs were not high on the agenda. In some ways, these unfortunate attitudes persist into the present (Charles 1995).

Second, the lack of an interdisciplinary, integrated view of fisheries has led to the dual tendencies of managers and policy-makers to (*a*) ignore the contributions of (non-economics) social science to fisheries thinking, thereby failing to appreciate the social costs of their actions, and (*b*) lack a vision of fishery systems as a whole and the complex interactions among its components, thereby failing to understand the ecological effects of management actions.

Third, in Canada's present mixed market economy, there is a great reliance on the price system in decision making by individuals. For example, price signals are meant to communicate to individual decision makers the societal cost and societal value of resources used in different ways. High prices supposedly signal scarcity, and provide the basis for decisions on current resource allocations such as the mix of inputs going into fishing effort. More general accounting and reporting mechanisms are also widely used, but in many cases need much further development.

Supposedly, prices are meant to guide decisions about the acquisition of both consumer goods (to be consumed relatively quickly) and assets (real property or investment goods, to be used over a relatively long time period). However, a central structural problem in fisheries management is the fact that there are prices to signal the value of fish as consumer goods, but there are no corresponding prices for valuing wild fish as assets capable of yielding future earnings (as breeding stock generating a future flow of fish to market). Hence conservation decisions cannot be made directly on the basis of a comparison of the value of fish as harvest versus the value of the fish in the sea. There is also no value placed in the price system on maintaining the integrity of coastal communities as a source of future productive capacity and component of social capital.

Another failure of the conventional price system lies in how it values the future relative to the present. A particular price of special interest in decisions involving reallocations over time is the "price of money" — the interest rate — which signals the productivity of investments, or the return which can currently be earned by foregoing current consumption or harvest. Market-based interest rates provide signals to investors as to the rate at which future returns should be discounted in market calculations in order to compare future returns with present costs and thus signal the expected return on investment (i.e., a return in excess of safe, low-interest instruments like bank accounts). However, the market-determined interest rate is usually thought to be significantly higher than the appropriate social discount rates by which society trades off future and present returns. In other words, the 'price of money' determined in the market for capital investments does not reflect society's preferences about the timing of consumption. Decisions based on this "price" will not properly incorporate social or ecological costs.

On the accounting side, similar problems exist. Environmental accounting and other adjustments of records of asset values are essential in business decisions, or more broadly, in social appraisals of resource management. A key gap in current management and social decisions is the failure to account for stocks of natural capital such as the value of a fish stock as an asset that can yield a future stream of scarce food. More generally, reporting on progress toward sustainability, taking into account all facets of human activities within fisheries, is essential in any assessment of broad social policy and social orientation.

As just noted, then, both market principles and informed social decision making demand that the consequences of a decision be fully evaluated in the sense that the value of resources consumed or given up be balanced against the benefits produced. Decisions will be distorted in cases where valuable stocks of fish are unpriced and hence ignored in decisions. Similar distortions will result where the value of alternative uses of resources consumed are overlooked, or where impacts on others (whether these impacts are injurious or beneficial) flowing from a decision are unpriced or unmarketed and as a result go unnoticed in the decision process. In an ecosystem perspective, many ecological services that have so far not been recognized as having commercial value and that therefore have not commanded any market price in trading, must be recognized as an element of valuable natural capital playing an essential role in ecosystem dynamics. Similarly, an important point throughout this report has been the observation that unpriced and unrecorded social capital, including human institutions and coastal communities, represents an asset which is crucial to productive use of physical and human capital, and profitable use of financial capital, but which is not traded or priced.

These various failures to account for social and ecological consequences of fishery decisions arise in the course of both societal and individual decision making. At the societal level, this may be accidental (for example, if fishery managers lack a comprehensive, integrated perspective on fishery systems) or it may be purposeful (if for example industry is deliberately pursuing narrow economic goals without regard for ecological or social consequences).

At the individual level, such failures are due in many cases to the absence of guiding prices on ecological and social flows and stocks. This can lead to decisions on harvesting levels and technologies which, though perhaps characterized as "efficient" in terms of market prices, are fundamentally distorted and wasteful because they presume zero value for crucial social and natural resources. (One attempt to estimate the scale of ecological services and stocks, priced or not, which can be identified as important to human activity can be found in Costanza et al. 1997.)

There is an urgent need to develop improved mechanisms to take into account social and ecological consequences of fishing and fishery management measures. Initiatives in this direction have occurred both at the market level and at the societal decision-making level. With respect to the former, a large number of studies, beginning with classic work by the welfare economist A.C. Pigou (1920), have now argued the need to improve market mechanisms by providing some estimates for unpriced services or impacts (see Hawken 1993; von Weizsacker 1994; Pearce et al. 1989). Some analysts (e.g., Knetsch 1996) have argued that, even in the absence of any methodology to assign plausible prices, decisions would be improved by forcing consideration of such unpriced flows through the assignment of arbitrary threshold values in place of the implicit price of zero. Indeed, in an ecosystem context, with the record of technological progress leading to the successive commercialization of species treated previously as waste, no component of the ecosystem is likely to warrant a value of zero. Waste of any such as yet non-commercial species or component is likely to prove in hindsight to be an unforgivable cost imposed on future generations. Similarly, the loss of traditional human institutions or the decay of coastal communities surely represent very real costs. No such institution or community deserves a value of zero, yet that is precisely the implicit value placed on these in much past decision making.

Turning to the level of societal decision making, proper social assessment of decisions demands adequate accounting for ecological and social services and assets. Since conventional accounting principles evolved for purposes of reporting on the affairs of commercial corporations, they are largely confined to the reporting of the value of marketable assets. When it is recognized that unpriced and non-marketed social and natural capital are key assets, then difficult adjustments to corporate and national accounts will be essential to reveal properly the true social impacts of individual, corporate, or social decisions. Again a vast literature has developed to address this question of environmental accounting (e.g., Repetto et al. 1989; Daly and Townsend 1993; van Dieren 1995) and social accounting (Waring 1988).

What happens if we do not take social and ecological consequences into account?

Failing to take all such ecological and social costs into account will typically lead to less desirable decisions. Indeed, many past actions (harvest levels, boat-building subsidies, reliance on quota controls, etc.) would probably have been quite different had social and ecological costs been properly considered.

A well-known example of the problem of not taking into account the full cost of fishing activities is the enormous worldwide loss associated with by-catch or discards of unwanted fish. Alverson et al.'s (1994) FAO report showed that in many fisheries, discarded fish were equal to half or more of the tonnage of fish landed. If the social and ecological costs of such wastage had been taken into account, fisheries regulations and evaluations of their success might have been different.

When social and ecological consequences are not taken fully into account in private and public decisions, we see exactly the fundamental distortions and errors in decisions that we currently are experiencing in fisheries systems: individual decisions are made without due attention to unpriced impacts on other participants in a given fishery (or often outside it); aggregate harvesting plans do not assign sufficient value to the resources wasted through by-catch and discards; and technological choices are made on the basis of "efficiency" calculations which assign no value to the resources wasted through by-catch or potential habitat destruction.

When the prices do not exist, or the accounting is not done, "externalities" (unrecorded costs) are not taken into account, the appraisal of potential investments is incomplete, the loss of natural capital is not reflected in estimates of economic growth, which consequently is overestimated, and the social costs of private economic returns are not examined.

More generally, if an adequate public assessment of government and social performance is to be possible, reporting should be structured to offer a balanced perspective on community progress toward sustainability in all dimensions of the natural and human systems. This entails going beyond simply economic, social, or environmental accounting to the construction of an integrated analysis assessing the well-being of both the natural and human systems in a balanced manner.

Principle 6. Clearly define roles, rights, and responsibilities of all fishery participants to align their interests with overall objectives of sustainability

> ***Principle 6.*** *Clearly define roles, rights, and responsibilities of all fishery participants to align their interests with overall objectives of sustainability.* We need to create institutions and incentive structures so that users of fish stocks are induced to maintain the resource in the long term and more generally so that individual behavior is compatible with the societal objective of sustainability. All stakeholders must be accountable first for the maintenance of viable marine and aquatic ecosystems and subsystems and consequently for the health of the fishing communities and enterprises that depend on the resource. Ultimately, a stewardship ethic must be fostered where it has the best chance of flourishing.

Justification

We have not yet sufficiently discussed the context and institutional setting within which decisions are made by people in fishery systems. Chapters 2, 3, and 4 emphasized the challenges of managing human activity in the complex natural and human structures which make up fishery systems. As well, the previous five principles have suggested ways in which management decisions in all components of these systems, public or private, might better take into account the variable, changing, and complex structure of the systems, and the risks and consequences associated with uncertain outcomes. We have emphasized an overall precautionary approach, methods for risk assessment, effective risk communication, and risk management, enhancing diversity throughout the system, developing more complete ongoing monitoring and information systems, and more accurately representing social and ecological consequences in individual and collective decision making. However, it is also necessary to examine the institutional context within which such decisions are made.

Specifically, previous chapters have emphasized the importance of collective action (to manage public resources in the public interest) and the importance of the incentive systems (to shape individual decisions and action in the direction of meeting societal goals). We noted in particular the ways in which inadequate or inappropriate regulation of access to common pool fishery resources may lead to problems of overinvestment in harvesting capacity, underinvestment in the well-being of coastal communities, and underinvestment in the future of fish stocks. This can lead to suboptimal social and economic benefits through dissipation of the potential economic return to the public owner (as a result of inefficiencies in harvesting and processing) or risks of virtual extinction of stocks or species as a commercial resource (as a result of overexploitation and habitat damage).

The fundamental issue here is that of designing (or recreating) institutions that help to create a consensus among participants and that avoid the problem of inappropriate or "perverse incentives." The latter problem is created when the rational decisions of individuals pursuing their own interests lead inexorably to outcomes that cannot be in the interests of all collectively. Such decisions are certainly not in the interests of the public owner of a resource that ought in principle to be renewable in perpetuity.

There is now a vast literature that addresses this problem of institutional arrangements and incentive systems that serve to align individual self-interest with an overall public or social interest (March and Olsen 1989; Ostrom 1990; Ayres and Braithwaite 1992; Lichbach 1996). One such mechanism is the "invisible hand" of market mechanisms described by Adam Smith (1776). However, before we leap to the solutions it offers, we need to be conscious of the problems of individual and collective actions associated with common pool resources, the requirements of a public owner determined to assure the sustainability of a heritage of a complex multispecies network of natural capital, and the concerns of a society that wishes to assure goals of social stability and equity in income and wealth distribution, which are of no interest to the "invisible hand." We also need to be aware of the extensive array of alternatives to the invisible hand mechanism which have been well-documented by social scientists in the last 15 years, although they are still largely unknown to policymakers (e.g., NRC 1986; Ostrom 1990; Jentoft and McCay 1995; Pinkerton and Weinstein 1995; Hanna and Munasinghe 1995). Worldwide case studies of contemporary non-market institutional arrangements for the management of common pool resources, including fisheries, have been rapidly accumulating and now include a substantial body of literature being catalogued at Indiana University (Hess 1996, 1998) and analyzed by scholars worldwide, especially through the conferences of the International Association for the study of Common Property.

Conclusions from this growing literature, in addition to an established public policy literature, are that we can and should design institutional arrangements that can effectively pursue society's objectives by limiting access to fisheries resources and harnessing individual motives of self- or public-interest. We need to do this in a manner that assures the following:

(*a*) ecological sustainability, based on primacy for conservation efforts, (through adequate constraints on harvesting efforts) and maintaining the health of future stocks, (through adequate continuing "investment" in spawning stocks, habitat preservation, and enhancement);

(*b*) viable industry and communities, based on harvesting, processing, and distribution activities, that avoid adverse impacts on fisheries, on other species, or on common habitat (unless compensated for in an ecologically acceptable manner);

(*c*) benefits to the public owner, in the form of whatever societal objectives are being pursued (profits, employment, equity, healthy coastal communities, etc.);

(*d*) competitive markets and social stability, through reasonable limits on concentration in industrial structure, wealth, and income distribution, and attention to the well-being of coastal communities;

(*e*) individual commitment to overall system goals through the exercise of individual discretion in changing and unpredictable individual circumstances, and through allegiance to agreed-upon rules and codes of conduct.

Chapter 6 outlines particular strategies for institutional designs to achieve these goals. They emphasize structures that encourage collective action (e.g., community-based approaches) as well as individual discretionary action consistent with overall design intentions and objectives (what we will call "compliance" for short) in the face of unexpected shocks and surprises in changing circumstances.

Of course, in order to make Principle 6 work, the "...overall objectives of sustainability" must be very well-defined. Throughout this report, we have generally stated these in terms of ecological sustainability, economic viability, and social stability. But clearly, in order to ensure that the incentives discussed here are properly "aligned," it is necessary to be much clearer about objectives. We do not propose specific details here, but rather briefly mention the process by which such objectives should be defined and communicated.

The type of participatory discussions that we have described previously can serve to enlighten all actors in fisheries about detailed objectives held by fishermen, coastal communities, processors, and government managers, for example. There will undoubtedly be a subset of objectives on which everyone can agree (e.g., need for long-term productivity of the fish populations on which economic and social benefits depend). The communication process will also make everyone aware of the objectives of other groups, so that while they might not agree with them, at least they will understand them. This could create greater compliance with regulations, for instance, if fishermen understand very explicitly that a certain regulation is not arbitrary but rather is designed to meet some conservation goal of reducing the probability of a stock collapse. Finally, to ensure that discussions about objectives are sufficiently clear, objectives should be documented in terms that are very specific, measurable (to know how close they are to reaching them), acceptable, realistic, and time-bounded (i.e., the goal should be reachable within a certain period).

Institutions are, however, a means to achieving good management, not an end in themselves. Many fisheries analysts have pointed to the fact that we need nothing less than fundamental change in human *attitudes* as well as social structures (Lawson 1993; Alverson and Larkin 1993). The seemingly perfect institutional design is just another structure if it does not in fact cause people to develop attitudes and values of stewardship toward the resource. Values of stewardship are not based solely or even principally on individual incentives. They are based on a conception of the social good being inextricably linked to the health of the resource. Among aboriginal peoples, this is, or was, usually expressed as spiritual belief. It is for this reason that we have some optimism about the potential of local-level institutions that are built upon, and that in turn foster, attitudes of stewardship toward local resources through a sense of basic connectedness. The environmental awareness and values of society in general also have a role to play, as explored in Chapter 6.

What happens if we fail to create institutions in which incentive structures align individual interests with social objectives of sustainability?

Most fisheries have the problem that, due to inappropriate incentive structures, individuals take actions in their own interests in ways that do not necessarily mesh with broader objectives of sustainability (ecological, economic, or social). Numerous symptoms of this illness exist. We see misreporting of catches, excessive discards and by-catch, development of excess capacity, and speculation in the sale of licenses, etc.

But the existence of perverse incentives is not restricted to the fishing sector. For example, past managers may have faced a perverse incentive to focus their attention on the fishing industry, which may have contributed to fishery management not taking adequate account of the linkages between regulatory measures taken in a fishery and the well-being of coastal communities. Broadly, we see examples of overexploitation, policy failure, inappropriate regulations given the risks, and reduced abundance and productivity of many fish stocks.

These phenomena have led in turn to reduced employment and social decay in many coastal communities. Governments have responded with continuing subsidies (albeit to a declining industry). Without appropriate incentives, these symptoms will continue. With appropriate incentives, the symptoms won't disappear, but at least we should see them less often.

Chapter 6. Implementing Principles for Sustainable Fisheries

Introduction

In this chapter we bring the discussion down to specific strategies and institutional innovations. These strategies offer concrete possibilities for action to move toward the goal of sustainable fisheries by implementing the six-fold package of general principles set out in Chapter 5. For each strategy, we give specific examples where elements of the strategy have been employed. We also identify some outstanding questions demanding further work.

Strategy 1 suggests action to develop a working precautionary approach to recognizing and taking into account structural and dynamic complexities, and the resulting uncertainties, in decision-making in fisheries systems (Principles 1 and 2). Because the precautionary approach may itself be viewed as an overarching strategy reflecting an evolution in ethical perspectives on the conduct of human activities in complex systems (Macdonald 1995; O'Riordan and Cameron 1994), some of these recommendations for action could easily be redistributed to one of the other strategies detailed below. However, given the extensive work already undertaken by many researchers and reported by the FAO on this subject (FAO 1995*b*, 1995*c*) and the central role the precautionary approach plays in the 1995 UN Agreement on Conservation and Management of Straddling Fish Stocks and Highly Migratory Fish Stocks (UN 1995), we considered it useful to gather in one place in our document the key elements of this work related to our particular topic. A further review of some key elements of the precautionary approach is provided in Richards and Maquire (1998).

Strategy 2 recommends much more extensive use of methods of risk assessment and risk management to take into account uncertainties and risks arising from the complex character of fisheries systems (Principle 2). These methods hold the promise of providing summary interpretations of a vast array of relevant empirical evidence and scientific analysis. However, such interpretations inevitably leave a dilemma (indeed the central dilemma underlying attempts to apply the precautionary approach in practice), namely a difficult social tradeoff between a reduction in estimated current returns (distributed over many interested individuals and groups) on one hand and dimly understood risks to fisheries systems, ecosystem integrity, and people in the future, on the other. Strategy 2 offers some examples of innovations toward more explicit analysis and open, participatory processes where social and political judgements about the resolution of these dilemmas may be developed in closer association with processes of information gathering and scientific interpretation, in light of traditional knowledge wherever appropriate.

Principle 3, which emphasizes the need to conserve and enhance biological, economic, and social diversity, translates concretely into Strategy 3, calling for action to implement integrated "systems" thinking in fisheries and in particular to implement a comprehensive ecosystem approach to fisheries management. A critical feature of this effort is assurance of adequate provision for the collection and communication of information relevant to fisheries systems as a whole (Principle 4) and this principle is carried forward into proposed concrete measures (Strategy 4) for the commitment of resources to particular initiatives to build the necessary capacity for monitoring and information management.

However, it is not just within fisheries systems that information is needed. To assess the performance of all those involved in activities affecting marine resources and to appraise the overall social significance of developments in fisheries systems, it is essential that the social and ecological consequences of decisions and actions be taken into account as fully as possible (Principle 5). Strategy 5 spells out some of the measures that are essential in order to respect this principle and offers some examples where the need for such significant departures from conventional practice has been recognized and some initial steps have been taken. Nowhere, however, is there sufficient attention yet paid to this pressing need for more adequate social reporting on progress toward sustainability (including ecological and social aspects) rather than the continuing reliance on the limited financial and economic statistics and (partial) stock data more routinely reported.

Finally, while steps toward a more appropriate focus on risk-averse policy and on an adaptive, ecosystem-based management approach more generally, are important and essential in the pursuit of sustainable fisheries, it is crucial that action not stop there. Fundamental to any hope of sustainability in fisheries systems is change in social attitudes toward use of natural and social capital and reform of the institutions governing decisions about access to and management of the resources of the global commons. Clarification of rights, roles, and responsibilities of all those involved is essential (Principle 6). Reform of the institutions of governance in fisheries systems at all levels, from local to global, is critical to ensure that institutional structures create appropriate incentives for individuals to pursue sustainable practices. The task is to align individual interests with the overarching objective of sustainable fisheries so that fisheries management policies and individual decisions all promote that social goal. Strategy 6 emphasizes features of such institutional arrangements and provides several concrete examples. Experience with some of these institutional innovations is recent and limited, however, and interpretations of the evidence differ. Conclusions as to which institutional forms will prove most effective in which circumstances are therefore contested and we do not put forward any single model as universally appropriate. What we do urge is attention to the issue and clarity in its resolution.

We believe that a key advance, and an important contribution of this report, is the growing recognition that it is essential to nest ideas about improved on-going current policy-making within a more general framework of adaptive management and open, adaptable, participatory management institutions and all of that within the yet more general frame of ethical evolution and institutional reform to *get the incentives right* by clarifying individual roles, rights, and responsibilities within more devolved, partnership-oriented institutions of governance promoting stewardship of marine resources and fisheries systems.

In particular, we develop later in this report the idea that the guidelines for the application of precautionary reference points laid out in the 1995 UN Agreement on Straddling Stocks and Highly Migratory Stocks (UN 1995 Annex II) may provide a basis on which ongoing management responsibilities can be effectively delegated within overall conservation limits established by government on the basis of scientific advice. With respect to management institutions, we see critical examination of a whole spectrum of possible arrangements spanning the range from *market* to *politics* as an important advance in debate about acceptable means to share and ration access to a resource that cannot be responsibly or sustainably managed as an open access commons. Examination of this range of options focuses attention in particular on the implications (merits, costs, and inherent risks) of transferability and trading of access or management rights (individual or collective) in fisheries systems. In the following sections we spell out these strategies in more detail.

Strategies

1. Implementing a precautionary approach

Principle 1 in Chapter 5 highlighted the need to recognize, in all decisions shaping human activity in fisheries, the underlying complexity of the systems involved. Principle 2 emphasized the need to take into account pervasive changes, uncertainties, and resulting risks. It is important that these principles be respected by everyone involved in fishery systems, i.e., people in industry and coastal communities, as well as those in management agencies.

The most appropriate way to build these two principles into our ongoing actions is to implement a precautionary approach. Such an approach aims to maintain productive fish stocks over the long term and ultimately to improve the social and economic benefits of fishing. It may, but will not necessarily, include actions such as reducing harvest rates, harvesting fewer non-target fish (reducing by-catch), or establishing areas that are closed to fishing (marine protected areas). While these actions may reduce economic returns *in the short term* compared to current amounts, these actions should be viewed as wise investments that will help ensure substantial, sustainable harvests in the long run. Such a long-term focus is particularly important given that some changes, such as loss of stocks and coastal communities, are irreversible, or only slowly reversible on a human time scale.

Many strategies are available for implementing a precautionary approach, but they are not mutually exclusive; a mix of strategies should be applied, depending on the particular situation. Several measures recommended below are either drawn directly from, or modified from, two important recent documents from FAO, the "Precautionary Approach to Fisheries, Part 1" (FAO 1995*c*) and "The Code of Conduct for Responsible Fisheries" (FAO 1995*d*). Many have been incorporated into the 1995 UN Agreement already cited (UN 1995).

Recommended strategies for implementing a precautionary approach are organized below under the six general characteristics of the approach, as described in Chapter 5. After making these recommendations, we provide evidence from international case examples to show that they can be helpful.

Characteristic 1. Unavoidable uncertainties (arising from natural variability and measurement error) should be recognized and explicitly considered when governments and management agencies evaluate regulatory policy options and when the fishing industry makes investment or other decisions. Greater caution should be exercised when information is uncertain, unreliable, or inadequate (UN 1995).

Recommendation 1.1. Although management agencies should use the most comprehensive scientific evidence available to make decisions, "the absence of adequate scientific information should *not* be used as a reason for postponing or failing to take measures to conserve target species, associated or dependent species, and non-target species and their environment" (Article 7.5.1 of FAO 1995*d*). The reason for this recommendation is that failure to take some restrictive action early because of uncertainty about its effect has often led to maintenance of the status quo. If that status quo happened to entail too high a harvest rate, then both the stock abundance and harvests subsequently decreased significantly (Saetersdal 1980).

Recommendation 1.2. The fishing industry and fish-dependent communities should make investment decisions taking full account of the uncertainty inherent in forecasts of the future status of stocks, as well as in effects of management regulations. They should not expect constant economic benefits from even the best-managed fish stocks, at the very least because stocks are naturally variable. Neither should the industry or communities base their expectations and investments in new harvesting and processing equipment on the best or even the average catch in the last five years (or some other period) because these may or may not be repeatable. Rather, they should be prepared to respond to changes in availability of fish at the spatial and temporal scale exhibited by the fish, even if those changes are at locations or times that do not fit ideally with human economic and social goals. Such biological variation is an unavoidable reality.

Recommendation 1.3. In the presence of climatic, economic, or technological change, it is particularly important that management agencies frequently re-estimate characteristics of fish stocks and fishing fleets, re-evaluate the appropriateness of current regulatory measures, and change those measures as necessary.

Recommendation 1.4. Fishery management agencies should design harvesting and other strategies to be robust, i.e., still likely to meet objectives even if the current understanding of the fisheries systems on which management decisions are based turns out to be incorrect, and (or) the capability to control fishing activity proves imperfect. While this robustness is not easy to achieve, some fundamental approaches help to move in this direction. First, robust management measures minimize reliance on the most uncertain or unobservable variables in fisheries. For example, where stock biomass is highly uncertain and setting of quotas is therefore difficult, harvesting during only a short period limits the possible fishing mortality without needing to know the stock abundance precisely. This is a common approach to in-season Pacific salmon management. Similarly, habitat protection, closed spawning and juvenile areas, and marine protected areas can help maintain food supplies, improve the age structure, and maintain the reproductive potential of the resources (Fisheries Resource Conservation Council (FRCC) 1997).

Harvest plans should aim at maintaining, to the extent possible, a typical proportional age distribution of the harvested fish stock to retain its ability to respond adequately to unfavorable environmental conditions; this may require minimal harvesting of prereproductive fish, or other size-selectivity of fishing gear, and a proportional harvest rate.

Management agencies should use a portfolio of management measures because no single harvesting or management method is likely to be perfectly robust. There will always be some situation in which any given method will fail to ensure sustainability. For example, the existence of a marine protected area (MPA) implies concentration of harvesting in other areas. Suppose an extreme case in which 50% of the fishing grounds are completely closed to harvesting. This might seem to guarantee protection for much of the fish stock. However, if the stock structure is uncertain, it is possible that such an arrangement, by concentrating harvesting pressure in the remaining open area, could wipe out substocks that happened to be located in that area. Thus MPAs should be used in conjunction with overall effort and harvest controls, as well as the spreading or rotation of fishing among areas.

Decision processes should emphasize adaptive management. Adaptive management involves a continuous monitoring of the fish resource and evaluation of past management actions, combined with a capability and willingness to make appropriate adjustments to those actions *in-season* and between years to ensure conservation. It is a fundamental tool for robust management. While many fisheries, such as those based on quotas, rarely see in-season adjustments, some others are managed on a frequently updated basis. In salmon fisheries, for example, abundance estimates are updated as the season progresses, with fishing times adjusted accordingly within the season. There is also some potential in combining quota controls with effort controls in marine fisheries, to provide a double-check that conservation measures are not thwarted. For example, if the overall effort limit is reached before the quota is caught, this may indicate that the quota had been set excessively high, thereby triggering a temporary closure of the fishery as a precautionary move.

Characteristic 2. Management agencies and the fishing industry should act to reduce the magnitude of uncertainties in the estimates on which decisions are based.

Recommendation 1.5. Management agencies should continue to improve methods of stock assessment, which produce estimates of stock abundance and productivity. As well, mechanisms for tracking changes in technological efficiency of fishing gear should be established. These will help improve forecasts of the effects of proposed regulations.

Recommendation 1.6. Due to confounding caused by simultaneous changes in the environment, management regulations, fishing gear, and fishing effort, agencies should implement management actions in an experimentally designed manner to permit a clearer indication of the effectiveness of particular regulations (McAllister and Peterman 1992; Sainsbury et al. 1997) and to maximize the learning opportunities from alternative courses of action.

Recommendation 1.7. As emphasized also in the discussion of Strategy 4, members of communities and the fishing industry should play a full role in helping managers obtain accurate statistics. Without these, estimates of abundance will be biased and a stock's productivity could be overestimated or underestimated, further increasing the uncertainties about the true status of a stock and its productivity and possibly increasing the chance of overharvesting, with consequent reductions in future revenue. Therefore, it is in the interest of the fishing industry to provide accurate data. To the extent that uncertainties are increased by industry's behavior or misreporting of catch statistics, management agencies may have little choice but to become even more precautionary in their regulations.

Characteristic 3. To maintain benefits from fishing over the long term, "where the likely impact of resource use is uncertain, priority should be given by everyone [particularly by management agencies and industry] to conserving the productive capacity of the resource" (FAO 1995*c*, p. 4).

Recommendation 1.8. "When there is a good year-class, give priority to using the recruits to rebuild the [spawning] stock rather than increasing the allowable harvest." (Article 48d of FAO 1995*c*). In other words, where harvests have been reduced or eliminated to allow a stock to rebuild, strong restrictions on harvests should be maintained for some time even after early increases in abundance are estimated. Among other things, this will increase the number of age-classes present in the fish population, thereby providing a buffer against unfavorable environmental variation (Chapter 3). This will help encourage the permanent recovery of the stock. If instead, successful cohorts from a depleted population are harvested too heavily too soon, this may prevent recovery to a high abundance, as was seen in many sardine and herring stocks in the 1960s and 1970s. The debate over the reopening of cod fisheries on the east coast of Canada must be evaluated in the context of this recommendation, i.e., do the potential benefits of economic revenue and improved information on the distribution and abundance of the stock outweigh the possibility that the stock could be prevented from recovering as a result of the openings? The issue is not straightforward. Recent (1998) evidence from catch rates, sentinel fisheries, acoustic results, and casual anecdotal observation suggests strong abundance inshore; however, the research vessel survey differs, suggesting a low overall stock abundance, with still depleted stocks offshore. Under the circumstances, what is a reasonable precautionary posture in the face of pressures from fishing communities, which may be giving unduly heavy weight to in-shore observations?

Recommendation 1.9. Despite apparent adverse direct impacts on current revenues, industry should support stringent restrictions on gear types or mesh sizes to reduce the detrimental effect of harvesting on the survival rate of juveniles and to minimize the shift in the age structure of the fish population that normally accompanies fishing. A diverse multiple-age structure of fish populations is important for maintaining the productive capacity of a fish population because it helps to compensate for spatial and temporal variability in environmental and human-induced factors that detrimentally affect fish stocks (the bet-hedging life history strategy referred to earlier in Chapter 3). Especially when uncertainties or natural variability are large, restrictive harvesting actions or limits to habitat disturbance will occasionally be necessary to leave sufficient spawning or rearing capacity so that fish stocks can successfully endure adverse environmental conditions and remain in a productive state over the long term. Such action is clearly in the long-term interest of those in industry and communities.

Characteristic 4. Steps should be taken to preserve and maintain the health of critical habitat.

Recommendation 1.10. Fisheries management agencies "should ensure that assessments of the implications of habitat disturbance are carried out prior to the introduction on a commercial scale of new fishing gear, methods and operations to an area." (Article 8.4.7 of FAO 1995*d*). Similarly, "In a precautionary approach to managing fishery technology, a designated lead authority should have the mandate to evaluate and decide on the acceptability of a proposed new technology, or changes to existing technology, and oversee the impact evaluation procedure." (Article 97 of FAO 1995*c*). This mandate should also apply to changes in the mix of technologies used, such as moving from mainly traps to mainly bottom trawls. Indeed, the impacts on habitat associated with all technologies in current use should be critically assessed.

Recommendation 1.11. "All critical fisheries habitats in marine and fresh water ecosystems, such as wetlands, mangroves, reefs, lagoons, nursery and spawning areas, should be protected and rehabilitated as far as possible and where necessary. Particular effort should be made to protect such habitats from destruction, degradation, pollution and other significant impacts resulting from human activities that threaten the health and viability of the fishery resources." (Article 6.8 of FAO 1995*d*).

Recommendation 1.12. Canada and relevant provinces "should ensure that their fisheries interests, including the need for conservation of the resources, are taken into account in the multiple uses of the coastal zone and are integrated into coastal area management, planning and development." (Article 6.9 of FAO 1995*d*). Such advice applies to all portions of watersheds in the case of Pacific salmon, not just the coastal zone, because of the importance of freshwater habitat for salmon.

Characteristic 5. The precautionary approach also requires that "harvesting and processing capacity should be commensurate with estimated sustainable levels of the resource, and that increases in capacity should be further constrained when resource productivity is highly uncertain." (FAO 1995*c*). While regulations to control the actual fishing effort during openings are key to an effective precautionary approach, reductions or at least constraints on the development of harvesting and processing capacity are also crucial because if that capacity were smaller, there might be less pressure on managers to open fisheries when biological information suggests that they should be closed (Botsford et al. 1997).

Although the pursuit of robustness is usually considered in the context of fishery harvesting plans (as in Recommendation 1.4), there are also fundamental policy measures that contribute to the above goal. Given that a highly uncertain resource must be harvested at lower mortality rates than would otherwise be the case and given that management controls are always imperfect, it is important to do whatever is possible to avoid the pressure that often arises from industry or among fishermen for increased harvests. Thus, robustness may be enhanced by some or all of the following:

- ensuring that catching power is compatible with the available harvest;
- ensuring that the benefits to society per fish caught are as great as possible, within the constraints of an integrated global marketplace, so that income levels are maintained while killing fewer fish;
- ensuring that the flow of information is as rapid and complete as possible, through local decision-making by those most familiar with, and most attached to, the fish resources and their environment; and
- ensuring that measures are taken to develop livelihood alternatives outside the fishery and to encourage fishermen to have outside employment, avoiding total reliance on fishing (thereby making a precautionary approach more acceptable to implement).

Recommendation 1.13. In all fisheries, regardless of whether they are new or existing, the default option should be that harvesting capacity and harvest rate should *not* be allowed to increase until quantitative analyses in a management plan (see Recommendation 1.16 below) show that the resource can adequately handle such increases. Agencies should be particularly careful to focus on limiting *harvesting capacity*, not just easily measured units such as numbers of fishing licenses. Harvesting capacity is determined not only by numbers of vessels, but also the gear used, the size of their holds, the volume of water searched and fished per unit time (which includes the effect of increasingly sophisticated electronic searching gear), and time spent emptying and preparing fishing gear for another episode of fishing.

Recommendation 1.14. Management regulations should use appropriate safety margins (reductions in harvest rates) that reflect uncertainties and risk. In the absence of a full analysis, these may be chosen arbitrarily (e.g., a 20% reduction in harvest rate) as a rough rule of thumb to take uncertainties into account. Preferably they should be determined by situation-specific quantitative analyses of risks and tradeoffs as outlined in the discussion of Strategy 2, "Extending risk-assessment and risk-management procedures."

Recommendation 1.15. Canada "should cooperate [with other fishing nations] to develop and apply technologies, materials and operational methods that minimize the loss of fishing gear and the ghost fishing effects of lost or abandoned fishing gear." (Article 8.4.6 of FAO 1995*d*).

Characteristic 6. Management plans should evaluate management options using the most comprehensive available methods and data, prior to the onset of fishing. Undesirable outcomes of management actions should be identified at that time, along with appropriate indicators, so that "any necessary corrective [regulatory] measures are initiated without delay, and … should achieve their purpose promptly" (FAO 1995*c*). These concepts apply to existing fisheries, newly proposed fisheries, and planned reopenings of fisheries that have been closed.

Recommendation 1.16. Management plans should include (*a*) specific management objectives, (*b*) consideration of a wide range of management options, (*c*) consideration of a long time frame (two to three decades), (*d*) explicit consideration of uncertainty when evaluating management options, (*e*) precautionary elements as defined in this section, (*f*) *pre-agreed decision rules* that specify *in advance* appropriate regulations for each of several potential circumstances so that action can be taken with minimum delay or debate, (*g*) mechanisms to monitor and control harvesting capacity and harvest taken, (*h*) input from industry, conservation groups, and communities, and (*i*) coordination with integrated coastal-area management plans (FAO 1995*c*). Management plans are particularly important for new fisheries and for cases where closed fisheries are being considered for re-opening. In both cases, the default regulation should be "no fishing" until a properly evaluated and designed management plan is approved.

Recommendation 1.17. "All fishing activities must have prior management authorization and be subject to periodic review" (FAO 1995*c*). Such authorization would potentially result from a management plan for a particular fishery. This option prevents the introduction of new harvesting strategies or fishing gear without full analysis of relevant information. As noted above in Recommendation 1.1, complete certainty about the effects of such changes is never feasible. Nevertheless, the best possible analysis that takes the available information into account, including its uncertainties, should be a prerequisite to new activities.

Recommendation 1.18. Management plans should specify predetermined reference points (as described in Box 4.2 in Chapter 4) in stock abundance or other variables to determine whether increases or decreases in harvest rates should be implemented. For instance, if current abundance estimates are below some predefined minimum amount, no harvests should be permitted.

Recommendation 1.19. "If a natural phenomenon [such as a climate-driven factor] has a significant adverse impact on the status of living aquatic resources, [fishing nations] should adopt conservation and management measures on an emergency basis to ensure that fishing activity does not exacerbate such adverse impact. [Fishing nations] should also adopt such measures on an emergency basis where fishing activity presents a serious threat to the sustainability of such resources. Measures taken on an emergency basis should be temporary and should be based on the best scientific evidence available." (Article 7.5.5 of FAO 1995*d*). Thus, closing a fishery is one of several options in an emergency.

Recommendation 1.20. For all fisheries there should be "an established legal and institutional framework for fishery management within which management plans that implement the [characteristics of the precautionary approach] are instituted for each fishery" (FAO 1995*c*).

Recommendation 1.21. Where feasible, management plans should be developed through consultations including management agencies, industry, coastal communities, and other relevant stakeholders. This will increase the chances that regulations will be followed and that they will achieve their purpose.

Examples of implementing the precautionary approach

It is time to take seriously a precautionary approach in Canada. Its importance and feasibility have been demonstrated worldwide by widely diverse management agencies, researchers, non-governmental organizations (NGOs), some fishing industry groups, and local fish-dependent communities (e.g., FAO 1995*c*, 1995*d*; Mangel et al. 1996; Gordon and Munro 1996; Canada 1995*b*, response to the Fraser report's recommendations, which called for *risk-averse* management; and FRCC 1997, "A Groundfish Conservation Framework for Atlantic Canada," in which the precautionary approach is advocated as an essential element of groundfish management). Several fisheries in various parts of the world have moved toward becoming sustainable in part by taking at least some of the measures outlined above. Below are examples where one or more elements of the precautionary approach have been applied.

Atlantic striped bass (NOAA 1993). This example illustrates placing a priority on protecting spawning biomass while a stock recovers from depletion. Commercial landings and abundance of juvenile Atlantic striped bass declined steadily through the 1980s after peaking in the mid-1970s. In response, U.S. state and federal authorities cooperated in implementing a comprehensive rebuilding plan for this species. A dominant feature of the plan was an intensive monitoring program with a minimum required precision to understand the reasons for the decline of the stock and to monitor the progress of its recovery. Regulatory measures included size limits and closure of major spawning areas to fishing to protect juveniles and spawners from fishing mortality. Specific regulations were introduced to prohibit fishing a strong year-class in 1982. As a result of these actions, juvenile abundances began to increase in the late 1980s. However, to permit the stock to rebuild, the stringent regulations were relaxed only slightly. Future relaxations were contingent on continued increases in various measures of stock status. This strategy succeeded and the Atlantic striped bass population was declared "fully restored" in 1995.

Snow crab in Gulf of St. Lawrence (Hare and Dunn 1993; Loch et al. 1995). The example of the Gulf of St. Lawrence stock of snow crab illustrates possibilities for reducing uncertainty through collection of better information and setting cautious quotas as part of a precautionary approach. This stock was on the verge of collapse in the late 1980s after years of ineffective regulation and enforcement. In 1990, several new measures were introduced to allow the stock to rebuild, including closure of molting areas to fishing and adoption of quotas below the most conservative estimates of annual recruitment. Industry and management have cooperated closely to improve data collection and enforcement systems. The result was an apparent recovery of the spawning stock, increased understanding of the biology of snow crab, and improved stock assessments. However, current stock estimates have raised concerns that perhaps even these cautious quotas will prove insufficient to prevent serious risk of reductions in future abundance.

Norwegian spring-spawning herring (ICES 1993). The case of post-1983 Norwegian spring-spawning herring demonstrates the use of a comprehensive management plan. Norwegian authorities stopped all harvesting on this stock after it collapsed in 1970, with no future harvest to be permitted until a "substantial" increase in spawning stock biomass (SSB) and recruitment was observed. Harvesting was resumed in 1984 but at a very low rate (about 5% of SSB) after the SSB had increased to 500 000 tonnes from its low of 100 000 tonnes. This was only after a management plan was developed that suggested that this low fishing mortality rate would have minor impact on the stock. The stock continued to rebuild and the 1992 recruitment was the largest recorded since the collapse of the stock in 1970. Although SSB was anticipated to be close to 2.5 million tonnes in 1994, authorities recommended that the low harvest rates continue. Current plans call for the harvest rate to increase gradually until SSB reaches its estimated maximum sustainable yield level of 6 million tonnes.

Spiny and slipper lobster in the Northwestern Hawaiian Islands (NOAA 1996). These lobster stocks exemplify use of management measures that place priority on protecting spawning biomass. Lobster fishing began in 1977 but by the early 1990s the catch per unit effort (CPUE) in the commercial fishery had decreased dramatically. The Fisheries Management Plan was amended in 1992 to limit further entry into the fishery, impose a 6-month seasonal closure to protect spawning biomass, and establish a flexible quota that could be increased, reduced, or eliminated entirely depending on pre-season and in-season abundance estimates. CPUE data since 1991 suggest that these species are recovering.

King, Tanner, and snow crabs in the Bering Sea (NOAA 1996). Reduced harvesting of these crab species illustrates how a comprehensive management plan that incorporates effects on the rest of the ecosystem can modify regulations. In addition to being the target of an intensive trap fishery, these crab species are also subjected to significant bycatch and discarding by Bering Sea groundfish fisheries. To reduce the amount of bycatch and discard, Alaskan management agencies have set a restrictive quota on total groundfish catches and have imposed a bycatch limit on groundfish fisheries. The result is that catches of many groundfish species are curtailed before their quota is reached because their bycatch limit has been exceeded and many groundfish species in the Bering Sea are therefore considered to be "underutilized." This is an example where a precautionary approach forces the fishing industry to forego short-term benefits from increased groundfish quotas in favor of maintaining biodiversity and abundance of other species in the Bering Sea ecosystem.

Research priorities

These strategies for implementing the precautionary approach go a considerable way toward ensuring the viability of fish stocks and fish-dependent communities. However, there is a continuing need for further research on several topics. (*a*) To reduce the confounding among environmental changes, changes in management regulations, and variations in fishing gear and effort, agencies should put more effort into designing management actions in an experimental framework (Walters 1986). While a statistically ideal experimental design is rarely achievable, there are still many opportunities for improving the design of management actions to increase the chance of attributing a cause to an observed effect (McAllister and Peterman 1992). (*b*) Researchers should continue to develop empirically based models to evaluate management options. (*c*) Agencies should put considerable effort into improving methods of stock assessment to reduce uncertainties about current parameter values. At minimum, agencies should continually monitor and update parameters of productivity to identify appropriate management actions in the presence of changes driven by environmental mechanisms or human activities. (*d*) "Research on the environmental and social impacts of fishing gear and, in particular, on the impact of such gear on biodiversity and coastal fishing communities should be promoted." (Article 8.4.8 of FAO 1995*d*).

2. Extending risk-assessment and risk-management procedures

In addition to implementing the aspects of a precautionary approach outlined above, planners and decision-makers in fisheries management agencies, the industry, and coastal communities should also put in place procedures for carrying out risk assessments and using the resulting information routinely in their decision-making. This is the remaining key component of Principle 2 in Chapter 5, "Recognize and take into account changes, uncertainties, *and risks*."

As noted earlier, many fisheries exist within a physical, biological, economic, and social environment that is inherently variable and uncertain. Ignoring the risks created by this extensive and pervasive variability and uncertainty will inevitably lead to poor decisions that generate ecological and economic losses. It is therefore crucial that risks be explicitly considered by everyone involved in fishery systems.

We strongly suspect that information on biological, economic, and social risks resulting from natural variability and uncertainties has either not been conveyed sufficiently clearly by scientists to fishery managers or industry, or that managers, industry, and communities have not responded appropriately to that information because the associated risks were not well-understood. In either case, it is essential that scientists put considerable effort into communicating information on variability and uncertainties and their associated risks to the fishing industry, fish-dependent communities, and fisheries managers to facilitate appropriate planning (see Strategy 4). It is also incumbent upon the industry and communities, as key players in a precautionary approach, to act upon that information in an economically and socially responsible manner, to maintain the productivity of the biological resource on which economic and social benefits are based.

Recommendation 2.1. Planners and decision-makers in fisheries management agencies, the industry, and coastal communities should implement explicit procedures for assessing (estimating) the risks associated with particular management strategies or actions. Specifically, risks can be estimated when developing a management plan and evaluating options, using quantitative methods such as simulation modeling.

Measures of risk involve a combination of the probability of some event occurring and the magnitude of its effect, if it does occur. For instance, there may be a low probability of collapse of a herring stock if only 20 more purse seines than planned are allowed to fish the stock during a short opening; however, large loss in future catches if a collapse occurs might mean that the risk is unacceptably large. Applying risk assessment to fishery management has become increasingly common (see Shepherd 1991; Smith et al. 1993; and Rosenberg and Restrepo 1994). Use of simulation models that incorporate random processes provides probabilistic assessments of age- or size-specific population levels, determination of targets and limits to exploitation (and the uncertainty in each), and evaluations of the probability that these reference points or management objectives will be met by particular actions.

Both the harvesting and processing sectors of the fishing industry frequently make decisions that involve evaluations of risks (e.g., those caused by fluctuations in market price, availability of fish or markets for selling fish, or social disruptions due to changes in fishery regulations). It should therefore be a simple step for them also to take into account other risks normally only considered by the management agencies, such as biological variation in survival and growth rates, spatial distribution, etc. It is as important for industry to do this as for the management agencies because, in the long run, failure to do so will lead to greater risks from overharvesting. For instance, industry should treat the change in fish abundance and productivity caused by environmental variability, fishing, and other human activities that affect habitat as another component of uncertainty in development of their business plans. This would be similar to the way in which uncertainty about future prices per tonne of fish is often considered.

Decisions are also made by individual fishermen taking risks into account. For instance, rates of discarding by-catch or relatively low-value smaller-sized fish (high-grading) reflect explicit decisions by vessel operators to leave room in the hold for handling higher-valued fish that might possibly be caught on a given trip (Gillis et al. 1995*a*, 1995*b*). In the Pacific salmon fleet of British Columbia, individuals who wish to spread their risks across space in case stocks or catches are reduced in any one area can do so, but now they must buy licenses to fish in more than one region (i.e., engage in license "stacking"). In Pacific herring fisheries in B.C., some boat owners have agreements at the start of an opening to share profits among themselves as a hedge against catching little or nothing.

This incorporation of variability into the planning "mind set" is consistent with the suggestion (Holling and Meffé 1996) that for ecological, economic, and social systems to maintain their resilience (ability to absorb disturbance or change), we must accept the presence and valuable role of natural ecological variability, which generates mechanisms to cope with changes in all parts of fishery systems, natural and human.

Recommendation 2.2. Formal procedures for risk *management* must also be implemented to ensure that results from formal risk *assessments* will be explicitly and quantitatively taken into account during the decision-making process and not in some arbitrary manner. In this way, the degree of uncertainty and the resulting estimated risks will influence the choice of management strategies, thereby providing a link between risks and precautionary strategies (see Chapter 4).

Methods of decision analysis, coupled with simulation modeling, are ideally suited to this task (e.g., Keeney 1982; Morgan and Henrion 1990; Walters 1981; Frederick and Peterman 1995). Decision analysis has been used by decision-makers in the business sector for decades and is a well-accepted way of explicitly considering risks when making decisions. Applying such methods in fishery systems requires an explicit consideration of a wide range of alternative hypotheses (not just the single best hypothesis) about the biological, physical, or economic processes that can affect the outcome of particular management actions. Consequently, decisions made using these quantitative techniques are more robust in the presence of uncertainties (von Winterfeldt and Edwards 1986) than the traditional approach of assuming that the scientists' best estimate, or most likely hypothesis, is correct.

Many resource agencies in North America, Europe, New Zealand, and Australia have already adopted this approach of quantitatively taking risks into account in decision-making (Rosenberg et al. 1993). As well, international bodies (e.g., Food and Agriculture Organization of the United Nations, the International Council for Exploration of the Sea, the North Atlantic Fisheries Organization) provide a logical focal point for adoption of this strategy.

Recommendation 2.3. Members of the industry, coastal fish-dependent communities, and fisheries management agencies should work cooperatively in project teams to estimate risks, using common sets of data and methods. Such teams should also explore via decision analysis the expected performance of various proposed actions (e.g., quotas, investments in different types of fishing gear) by all of those sectors.

Institutional links must be designed so that processes for collecting, analysing and interpreting data, and using the resulting information in an integrated manner for decision purposes, are well-coordinated and provide for participation of a wide range of interests as well as differing scientific perspectives. (Recall Box 4.4, which introduced the concept of "civic science.") Scientific work and policy judgements within government management agencies could be supplemented more regularly both by independent scientific advice and the perspectives of participants in fisheries systems. Scientific information might include more extensive linkage with the academic community as well as advice grounded in basic and applied research such as was offered by the former Fisheries Research Board of Canada and now by regional Conservation Councils such as the Fisheries Resource Conservation Council (FRCC) on the East Coast, and perhaps the new Pacific Fisheries Resource Conservation Council, or the proposed Fisheries and Seafood Diversification Board in B.C.

Views of various participants in fisheries systems should include those based on local and traditional ecological knowledge and ideas grounded in community-based consultative processes as envisaged in the proposed joint federal–provincial Fisheries Renewal Advisory Board in B.C. (The design of institutional structures to best serve this process of carrying science effectively into the policy process is a difficult question that has been addressed to some extent in recent debate (Hutchings et al. 1997; Doubleday 1997). As discussed below, some members of this panel see real promise in the concept of precautionary reference points embodied in the recent UN Agreement (UN 1995) in which conservation, or limit, reference points are established by governments on the basis of independent scientific advice, while the development of management, or target, reference points is delegated to participatory management agencies, stakeholder groups, or co-management structures.

Recommendation 2.4. The application of the precautionary approach requires that the burden of proof be removed from scientists and managers, so that risk-neutral or risk-averse policies can replace the clearly demonstrated risk-prone strategies of the past. If this change is not effected, it is unlikely that the advantages of the combination of risk assessment with the precautionary approach will be achieved (Peterman 1990; Dayton 1998).

This recommendation results from the observation that the principal difficulty in implementing an approach that couples risk assessment with a precautionary approach is the issue of the burden of proof. A long-standing problem in national and international management has been an implied burden of proof resting with scientists and managers to demonstrate adverse impacts of harvesting on marine populations and ecosystems. This tradition represents a significant obstacle to responsible decisions. It has been extremely common to see the recognition of high levels of uncertainty used as a rationale not for appropriate caution, but rather for continued high levels of exploitation, i.e., as an excuse not to reduce fishing activity. This practice is not tenable.

Examples of applying risk assessment and risk management

Bergh and Butterworth (1987) describe how evaluators of management options for the South African anchovy considered risks by simulating the effects of recruitment variability and imprecision in estimates of abundance. Managers adopted a different fishing strategy than what would have been used had those risks been ignored.

Other examples are given in Kirkwood and Smith (1996), including references to recent International Whaling Commission (IWC) work on investigating management strategies for whaling that took uncertainties explicitly into account, as well as to the adoption by the Council for the Conservation of Antarctic Marine Living Resources (CCAMLR) of various aspects of the precautionary approach. Use of risk analysis by the FRCC in determining desirable TAC levels for Georges Bank cod and haddock fisheries may also be cited (FRCC 1997). This work focused on setting quotas to ensure a high probability of stock rebuilding. Huppert (1996) also provides some useful examples (see also Chapter 4).

Research priorities

The central research issues in the development of approaches to risk assessment involve the development of methods of quantifying uncertainty, particularly that associated with rare events, and in assessing possible responses to extreme events. There has been substantial interest in the development of Bayesian statistical approaches to provide probabilistic assessments of key population parameters and state variables in fisheries, although there is still controversy about the application of these methods. Further development and evaluation of this approach and recommendations for robust approaches for development of prior distributions would be extremely useful in a fishery risk-assessment context. Many risk-assessment approaches depend extensively on resampling techniques such as the bootstrap method to compute probability distributions of parameters and state variables (Smith et al. 1993). There are four general approaches to bootstrapping; the performance of each deserves comprehensive evaluation in this context.

Within the social science arena, there is need for more research on economic and particularly social risks of various actions. For instance, investigation into the response by fishermen to risk and their perception of risk would be extremely helpful. This would include finding ways to bridge the commonly observed gap between estimates of risks by analysts and perception of risks by others (Slovic 1987). As has been noted in other contexts (e.g., Moffet 1996; NAPA 1995), this task may entail extensive consultative processes and public participation exercises aimed at encouraging social learning about risks. In addition, current research in cognitive psychology shows that traditional scientific means of communicating information about uncertainty and risks are not necessarily the best for assuring understanding or comprehension by target audiences. Research into better methods should be encouraged and the scientists, analysts, and managers who should be using this research should be exposed to it (Morgan and Henrion 1990). Finally, a substantial element of implementation error (see Chapter 4) would be reduced if we could understand ahead of time the probable response of fishermen to changes in regulations. Increasing interest in institutional design and incentive structures as concerns for public administration, particularly in the field of implementation, is leading to relevant material in this area (see, for example, March and Olsen 1989). In this respect, work to improve methods used to communicate uncertainties and their implications to managers, users, the public, and scientists would be valuable in improving awareness of risk-management problems and in promoting compliance.

3. Promoting diversity

Principle 3 sets out the general argument for measures to promote and conserve diversity as a key feature of ecological health and as a hedge against uncertainty and risk in fisheries systems. A number of strategies offer opportunities to pursue this principle, including the adoption of an ecosystem-based, adaptive, and precautionary approach in decisions by all those involved in fisheries management, broadly interpreted to include government and non-government organizations, industry, and people.

Recommendation 3.1. Pursue an ecosystem approach. In the 1996 statement of goals and objectives for the Ecological Monitoring and Assessment Network (EMAN) in Canada, the ecosystem approach is defined as "a comprehensive and holistic assessment of the environment [which] recognizes the complexity of ecosystems, emphasizes that people are indeed part of ecosystems, and adopts ecozones as spatial reporting units. Such an approach examines the interconnections within and among component parts." (EMAN 1996).

Box 6.1. What is an ecosystem approach?

Among the most widely known illustrations of an ecosystem approach is the work of the International Joint Commission to implement the 1978 Great Lakes Water Quality Agreement. Slocombe (1993) suggests that "[a]s applied in the Great Lakes Basin, the ecosystem approach is holistic, interdisciplinary, goal-oriented, participatory, and aimed at getting people to recognize that they are part of the ecosystem not separate from it ..." Drawing on a wider set of examples, Slocombe identifies a set of core characteristics of ecosystem approaches as follows:

(*a*) describing parts, systems, environments, and their interactions;
(*b*) holistic, comprehensive, transdisciplinary;
(*c*) including people and their activities in the ecosystem;
(*d*) describing system dynamics, e.g., with concepts of homeostasis, feedbacks, cause-and-effect relationships, self-organization, etc.;
(*e*) defining the ecosystem naturally, e.g., bioregionally, instead of arbitrarily;
(*f*) looking at different levels or scales of system structure, process, and function;
(*g*) recognizing goals and taking an active management orientation;
(*h*) including actor–system dynamics and institutional factors in the analysis;
(*i*) using an anticipatory, flexible research and planning process;
(*j*) entailing an implicit or explicit ethics of quality, well-being, and integrity; and
(*k*) recognizing systemic limits to action, defining and seeking sustainability.

In a review of what he terms ecosystem management, Grumbine (1994, p. 31) identified ten dominant themes, which led him to a working definition: "Ecosystem management integrates scientific knowledge of ecological relationships within a complex socio-political and values framework toward the general goal of protecting native ecosystem integrity over the long term."

Here we emphasize, as in the previous discussion, the need for a multispecies approach to fisheries systems, with full recognition of the spatial and temporal heterogeneity of the component parts, including the human activities and institutions involved. Integrated assessment and planning processes that include terrestrial, coastal, and ocean processes, and integrated procedures for land use and marine resource use are essential as part of this broad ecosystem-based approach to management.

Such an approach will emphasize, in particular, the importance of interdependence in ecological networks and food webs (Arrow et al. 1995; Pauly et al. 1998).

Recommendation 3.2. Avoid impacts on the ocean habitat. (See also Recommendations 1.10–1.12). In many fisheries, concerns are expressed over the effect of harvesting technologies on the habitat and on fishery conservation in general. However, without complete knowledge of the ocean, the impact of such harvesting on both the fish and the habitat will always be uncertain. In particular, uncertainties in ocean ecology ensure that it will be difficult to prove (or disprove) negative impacts on the food chain and on ocean productivity. In this situation, harvesting is more robust (and more precautionary) if carried out using technologies with the least potential impact. For example, in seeking to be cautious in the 1997 reopening of fisheries for two Atlantic cod stocks (in the northern Gulf of St. Lawrence and southern Newfoundland, Areas 3Ps and 4RS3Pn), fishermen agreed that no trawlers would be used.

Recommendation 3.3. Promote marine protected areas. (See also Recommendations 1.4 and 1.11). One timely specific initiative is to create a network of marine protected areas to preserve spawning populations, nursery areas for juveniles, feeding grounds for certain life stages, and the integrity of the ecosystem in general (DFO 1997*b*). There are many reasons to prohibit consumptive human activity in certain parts of the ocean and to coordinate economic activities with conservation measures in others. If one is focusing on a particular fish stock, then designating spawning or nursery grounds as "closed areas" may make sense. More broadly, an ecosystem approach to fishery management implies that human impacts on all species, and on the ocean habitat more generally, must be taken into account.

Marine protected areas (MPAs, also referred to more or less synonymously as marine reserves or marine conservation areas) are areas of the ocean in which exploitation is prohibited or clearly limited. This is done to protect critical habitats, endangered species, or commercially harvested species that need "sanctuaries" within which they can grow and reproduce in the absence of harvesting pressure. Such protected areas create "buffers" that help a population persist in the face of environmental or human-induced variation (Gubbay 1995). In some species it may be possible to significantly reduce harvesting on entire subpopulations that may be genetically unique. These subpopulations can be held in reserve in case the main harvested population collapses. Such reserves thus constitute a bet-hedging strategy, similar to what is used by financial managers with diversified investment portfolios (Gordon and Munro 1996) and by long-lived fish species to help them survive variations in conditions over time and space.

Different designs are possible. Seasonal sanctuaries or refuges may be designated, for example, to protect spawning activities at certain times of the year. MPAs may be "zoned" for multiple use (to allow harvesting or other activities such as tourism in certain subsections of the MPA) or they may be fully protected "no take" areas (in which all exploitation is prohibited). The latter approach provides an important hedge against uncertainty: by leaving part of the ocean ecosystem untouched, it is possible that at least a segment of various populations at risk may be preserved. In this fashion the MPA may recreate the natural refuges that contributed to the sustainability of some stocks that successfully resisted heavy fishing pressure until recent technological developments rendered these sanctuaries accessible and open to exploitation.

MPAs are presently receiving a great deal of attention among researchers, as noted in the discussion of Recommendation 1.4. The excellent DFO discussion paper (DFO 1997*b*) outlines its proposed approach to the establishment and management of MPAs under the 1997 Oceans Act, extending and complementing the existing marine conservation programs of Canadian Heritage (National Marine Conservation Areas) and Environment Canada (for habitat and wildlife, especially migratory birds). That initiative deserves high priority.

Recommendation 3.4. Adopt an integrated approach to ocean and coastal management. Many coastal fisheries in Canada and globally face a trio of fundamental problems: overexploited resources, overextended fleets, and a lack of non-fishing alternatives. Addressing them requires integrating fishery management efforts and coastal economic diversification. Generating employment alternatives is crucial to effective resource conservation (Smith 1981). To this end, fishery planning can be combined with community-based diversification. For example, in combination with efforts to reduce fishing capacity, emphasis can be placed on creating employment alternatives that build economic strength within the community, taking advantage of local comparative advantages in ocean-related activity (such as development of alternative fisheries, fish farming, coastal tourism, etc.). There is a growing literature on this approach, referred to as "Integrated Coastal Development" (e.g., Arrizaga et al. 1989) or "integrated coastal zone management" (Wells and Ricketts 1994).

It is often argued that public policy has social as well as economic dimensions and these may call for investment in sustaining coastal communities even in the face of apparently conflicting economic pressures. What is less well-recognized is that this need not be an argument for creating or perpetuating economic inefficiency and social dependency, but simply acceptance of a broader notion of economic efficiency, which takes into account the value of unpriced social, cultural, and natural capital. For reasons of lifestyle and independence, people are often willing to exploit their own labor and human capital for incomes they find unacceptable as paid wages. At these lower income levels, which may be acceptable in small communities with substantial family and social networks, selective harvesting and habitat maintenance activities, which would otherwise not be commercially viable, become feasible. These activities, viable on a continuing basis in coastal communities, contribute unpriced social and ecological services, which may offset (more than fully) perceived subsidies or regulatory burdens required to ensure the survival of the communities. The appropriate social balancing of such policies can only be undertaken in the context of the integrated analysis just mentioned.

Examples

As noted above, perhaps the outstanding example of implementing an ecosystem approach lies in the work of the International Joint Commission (IJC) in the Great Lakes Basin. This work is oriented toward participation and coordination of decisions and activities of many institutions in the region and emphasizes the need for all to see clearly the role that they can play in decision processes. As in Chapter 4, the IJC underlines the need to deal explicitly both with uncertainties related to understanding a decision problem and uncertainties arising from the possibilities of error in policy decisions and in implementation of them. In the later risk-management phases of the work, the importance of broad participation and "civic science" (Box 4.4), along with a precautionary approach, is emphasized (IJC 1995).

Another example of a promising attempt to pursue a comprehensive ecosystem-based approach is found in the work of the Central Region Board in British Columbia to develop an integrated approach to terrestrial and marine resource use on the West Coast of Vancouver Island, building on the conceptual foundations provided by the Clayoquot Sound Scientific Panel (CSSP 1995). There is now widespread community and industry support for an application by Canada for designation of this region as a UNESCO Biosphere Reserve, possibly including important marine protected areas.

Lobster fishermen on the Eastport Peninsula of Newfoundland themselves closed two traditional lobster fishing grounds for conservation reasons starting in 1997. Examples of more formal efforts to create marine protected areas can be found in several places, including recent federal and British Columbia initiatives. (Indeed, the popularity of the idea is such that at least three pieces of recent legislation, i.e., the Oceans Act, the proposed new Fisheries Act, and provisions under the proposed Endangered Species Act, as well as federal and provincial legislation on parks, provide authority for such action.) DFO, through development of its "oceans strategy" (Department of Fisheries and Oceans Canada 1998*a*) aspires to ensure the necessary coordination of such initiatives.

Priorities for research and further work

The need for integrated systems research on fisheries and oceans matters has been highlighted in the United Nation's Agenda 21 document. In Canada, the FRCC recently emphasized the need for integrated fishery analysis (FRCC 1994), recommending that "a multidisciplinary team approach be implemented in addressing fishery research questions, both in the laboratory and in the field" and that "a real move be made towards an ecosystem approach to fisheries management." The FRCC also suggested that: "scientists study fishing scientifically as a system and strive to better understand the relationship between fish (resource) and fishing (fishing practices, gear technology, capacity analysis, etc.). This must reflect the recognition that fishery science involves more than the natural sciences and that scientific research is a part of the development, implementation and evaluation of fishery management measures and economic policy tools." (FRCC 1994, p. 118).

Such a research approach supplements rather than replaces disciplinary lines of research. Several relatively new multidisciplinary areas of research show promise for moving in this direction. These are described briefly below.

(*a*) Ecological economics

Ecological economics is a new field of study that attempts to integrate economic and ecological thinking, in order to put some analytical and conceptual flesh on the idea of "sustainable development." It recognizes the key roles of economic forces, of ecological constraints, and of societal objectives. Indeed, these components were highlighted in a recent overview of ecological economics, in which three major problem areas for study were noted: (*1*) assessing and insuring that the scale of human activities is ecologically sustainable; (*2*) distributing resources and property rights fairly, both within the current generation of humans and between this and future generations, and between humans and other species;

and (3) efficiently allocating resources as constrained and defined by 1 and 2 above, and including both marketed and non-marketed resources (Folke et al. 1994).

Ecological economics is systems-oriented and emphasizes the importance of natural capital (including fish, natural habitats, the oceans, etc.) and the ecological services (or ecological life-support) provided by nature. These ideas are nothing new in fisheries. For example, fish stock conservation is a matter of "maintaining natural capital," the marine ecosystem provides ecological services, and fishery management is a matter of balancing conservation (an investment in natural capital) with harvesting (conversion of natural capital into consumption flows). Concrete applications in fisheries policy remain to be pursued.

(*b*) Sustainability assessment

While discussions of "sustainable development" (the balancing act between present-day benefits and future rewards) have become commonplace over the past decade; only recently have widespread research efforts focused on measuring and evaluating the sustainability of human activities. Interestingly, however, such challenges have been at the forefront of fishery research for decades (Schaefer 1954; Beverton and Holt 1957; Gulland 1977; Ricker 1975). In the past, this fisheries research focused principally on achieving particular goals related to "sustainable yield" if not the maximum, then some value less than that based on other criteria such as aversion to risk. Measures of sustainability now need to consider effects of human activities on fisheries systems more broadly. In this spirit, the challenge can be envisioned as achieving simultaneously acceptable performance in the ecological, social, and economic components of fisheries systems. This will necessarily involve tradeoffs among these components and a proposed fishing activity or management regulation would be considered unacceptable if its impact on any one component would be excessively detrimental. Many possible combinations of conditions of these biological and human components might be sustainable in the long run; however, some combinations would be less acceptable than others, depending on the relative values and ethical weights assigned to the various ecosystem components. In many circumstances management efforts may also pursue restoration or enhancement of the ecosystem, rather than sustainability alone.

Examples of assessing and reporting on progress toward sustainability in this general sense are noted briefly in discussion of Strategy 5 below.

(*c*) Integrated fishery modeling

Integrated fishery modeling focuses on the need to understand the interrelationships among components of fisheries systems through multidisciplinary means. Efforts to develop integrated analyses are generally seen to date from the 1950s, progressing further into dynamic analysis in the late 1960s, notably with the predator–prey style models by Smith (1968) of fish–fishermen interactions. The fundamental ingredient of such models is a multidisciplinary combination of components of fisheries systems.

A major focus of integrated modeling has been in development of bio-economic models, combining fish population dynamics with economic dynamics and objectives (e.g., Clark 1985). Bio-economic modeling has enabled researchers to address the dynamics of both fish and fleets, while also providing a language that can help bridge the gap between biologists and economists working on common projects. There is considerable potential to utilize the approach on case studies using simulation modeling (e.g., with FAO's BEAM IV software). There have also been efforts to incorporate multiple objectives, as well as social and behavioral aspects, through "bio-socio-economic" approaches (Sivasubramaniam 1993). It should be noted that these more integrated approaches highlight the full range of data needed in fisheries, such as time series of labor forces, labor force participation rates, and fishing community populations, as well as the more usual data on fish stock dynamics and economic parameters.

Key goals of integrated fishery modeling include (*a*) an understanding of how fishing fleets, fishing effort, and labor change over time, (*b*) the ability to predict the responses of fishermen to regulations (such as the experience of the Atlantic groundfishery, in which restrictions on vessel length designed to limit capacity expansion, were met with construction of *wider* vessels, so that capacity expanded nonetheless), and (*c*) an understanding of the decision-making inherent in illegal fishing behavior and the consequent effectiveness of enforcement. Of course, until modeling efforts are better able to represent the multispecies, joint production (interdependent) network character of the ecosystem itself and the mechanisms generating ecological services, the potential for fully integrated modeling will be limited. Further research must be directed toward such representation.

4. Ensuring adequate collection and communication of information

Information is a key element in shaping human activity. Principle 4 in Chapter 5 emphasized the importance of measures to ensure a comprehensive continuing flow of information. Integrated fisheries modeling and a coherent ecosystem approach to management depend on it. An appropriate plan for its collection and dissemination is important in any endeavor but particularly so in fisheries, of which we have so little fundamental understanding. In the absence of a clear model, such as that which meteorologists use to make predictions for the atmosphere, it is very difficult to organize and understand the observations that we do have. It is, therefore, primarily through the observations of fish in their environment that we understand their status, and similarly for the fishermen and other stakeholders in their social environment. It is crucial that we devise a scheme for the integrated sampling of fisheries because of the strong positive and negative feedback between different components of these complex systems.

Presently, there are signs that the DFO has opened up the information flow and substantially improved communication within stock assessment procedures and further increases in transparency are proposed. For example, the steps now being taken to follow up the federal–provincial Memoranda of Understanding on setting up a Pacific Fisheries Resources Conservation Council and a Fisheries and Seafood Diversification Board in B.C. include extensive public consultation on structures, policies, and procedures. The advice offered to Ministers by both bodies is to be contained in reports that will be available to the public.

In addition the importance of timely and relevant information built into a precautionary approach to the development of new and emerging fisheries is recognized in the discussion draft Policy and Guidelines for the Development of New Fisheries in B.C. (Department of Fisheries and Oceans Canada 1998*b*). These draft policies follow the staged approach outlined in the Pacific Stock Assessment Review Committee (PSARC) Working Paper I96-06 (Perry et al. 1999) designed to assure an adequate information base to support decisions on commercialization of new fisheries or harvesting methods. The effective integration of traditional ecological knowledge for these purposes evidently will be a key concern as will be the application of effective methods to assess the impact of such developments on marine food webs and to force the necessary selectivity in harvesting technologies (Pauly et al. 1998).

Such changes should be encouraged. However, the information flow in this process is still restricted relative to the much broader approach that we recommend. Focus on the marine ecosystem, fish, and the environment, fails to take other sources of information fully into account. One possible first step in the process could be the establishment of an open, integrated, and freely accessible database. All departments, agencies, companies, and university researchers could contribute information for inclusion into the database. Such a collation would make it possible to carry out the integrative analysis required to understand some of the complexity of fisheries systems.

It is particularly important that social and economic streams be more fully and systematically added to the information flow. The benefit of their addition should be obvious to managers who have actually worked to implement a management plan. With appropriate awareness and understanding of the social and economic environment, a management strategy and plan can be crafted that maximizes the chances for acceptance and effectiveness.

In particular, we make the following specific recommendations:

Recommendation 4.1. Implement an integrated, balanced, and transparent information-flow strategy.

Recommendation 4.2. Involve fishermen and other groups in the collection of information.

Recommendation 4.3. Make better use of existing information resources including traditional knowledge and knowledge of local fishermen.

Recommendation 4.4. Involve those who collect information in the decision-making process.

Recommendation 4.5. Improve information flow so that it is open and accessible to everyone involved in fisheries systems.

Examples

On the Atlantic coast, the FRCC was established to provide the Minister of Fisheries and Oceans with advice for groundfish and other fish stocks. Among the achievements of the council are a more open and participatory communication process and more extensive and rapid information flow. On the other hand, the Council has operated under constraints, e.g., it has been largely restricted to dealing with groundfish, crab, and lobster, and with a limited mandate reducing its potential effectiveness. There is need for a system and process like this but on a scale broader in its geographic focus and in its coverage of species and ecosystems. We must bring together all the various players, turning them into partners in the search for understanding.

In the Atlantic groundfish fisheries, various measures have been instituted to develop a participatory approach to enhancing information collection and availability. A notable example is the Fishermen and Scientist Research Society, an organization of fishermen and scientists in Nova Scotia, who jointly collect and analyze biological, oceanographic, and catch data that would not be collected without their efforts. A similar participatory approach is used in the range of sentinel fisheries that take place in closed or newly reopened groundfish fisheries; these involve a combination of structured research surveys together with low-level harvesting to provide catch rate indices. All the activity is carried out by fishermen, under the scientific direction of DFO scientists. A third example is a survey carried out by trawler fishermen in Southwest Nova Scotia (NAFO Area 4X) that provides an independent stock status indicator, one which covers more inshore areas than is possible with DFO's offshore research vessel survey.

We argue that this participatory approach should be common to all fisheries. It should be seen as an ongoing monitoring requirement, not simply as a means to save money or provide support to unemployed fishermen.

On the Pacific Coast, agreement has been reached on the main features of the Pacific Fisheries Resource Conservation Council (PFRCC) to be implemented as a joint federal–provincial initiative. This body will have a Chair and six members appointed by the Council of Fisheries Ministers to offer advice on matters of conservation and long-term sustainable use. Regrettably however, it appears that the terms of reference will confine the PFRCC, at least initially, to salmon fisheries alone. It will not include representation of sectoral interests and will not deal with issues of allocation or in-season management. Rather, it will be expected to receive scientific information from the PSARC and more generally from other bodies, including First Nations, in order to offer strategic advice on habitat and stock conservation and enhancement in a more open, inclusive, public process than PSARC itself follows. The PFRCC will have an explicit obligation to achieve an integration of traditional ecological knowledge where possible and to assure data are publicly available and accessible. In particular, it is expected to make its recommendations available simultaneously to the Council of Fisheries Ministers and the public in January of each year. Thus the scientific advice from the existing PSARC process is to be brought into a more general process in which integrated strategic advice may bridge troublesome federal–provincial jurisdictional conflicts in responsibilities for management, habitat, and land-based activities more effectively.

Again on the Pacific coast, the West Coast Trollers Association, including about 300 members, was formed in 1996 from the trollers who selected Area G, the west coast of Vancouver Island, as the area in which they would exercise their trolling license privilege under the "Mifflin Plan" for reducing the west coast salmon fleet. This Association quickly identified the need to improve management through accurate and timely catch monitoring in-season, and after exploring the best available information on how to achieve this, they designed a project that they were willing to pay for out of their catch allocation (sale of fish "off the top"). The monitoring project would involve about 10% of the trollers, who would use on-board low-cost, purpose-designed data keyboards linked to a Geographic Positioning System (GPS) and satellite communications hardware. Observers would also be placed on a subsample of the vessels. The position of each boat would be recorded and linked to their call-in (four times daily) of catch and by-catch. Catch, CPUE, and catch location would be plotted daily on computer-generated maps at the Fisheries Centre at the University of British Columbia, enabling an analysis of the best selective fishing strategies (i.e., in what location can the fleet maximize the harvest of targeted species while minimizing by-catch). This information would be relayed to the West Coast Sustainability Association (the community-based association of West Coast trollers), to DFO, and the public about a week later, with vessel identification stripped from the public-access version. The West Coast Trollers Association aspires to be the first fleet in Canada to receive Marine Stewardship Council certification for a sustainably pursued fishery and continues to pursue this proposal as a way of testing different selective fishing strategies. At the time of writing, however, an implementation agreement had not yet been finalized with DFO. There are, however, joint data collection activities occurring among DFO and users in some invertebrate fisheries.

Although the use of traditional ecological knowledge is required under the new Canadian Environmental Assessment Act and a number of promising initiatives have been undertaken, we can point to only a few examples where the use of such knowledge is effectively integrated with conventional information flows in ongoing fisheries management. Arctic fisheries are now managed under co-management agreements with DFO that place high priority on the use of traditional knowledge. For example, this feature is built into the structure of the Nunavut Wildlife Management Board and the Fisheries Co-management Board created under the Inuvialuit Final Agreement. In Nova Scotia, scientist members of the

Fishermen and Scientist Research Society mentioned above regularly refer to local knowledge gained from their interactions with fishermen.

Research priorities

The ocean is such a vastly undersampled system that research is crucial to improve the understanding that can be gleaned from the little data that we are able to collect. Interdisciplinary research is beginning to blossom within the field of ocean ecology, witness the development of programs such as the Global Ocean Ecosystem Dynamics Program (GLOBEC), the Joint Global Ocean Flux Study (JGOFS), as well as the World Ocean Circulation Experiment (WOCE) and the Land-Ocean Interactions in the Coastal Zone (LOICZ). All have something to offer fisheries, although relatively few broad programs exist that are focused on fisheries as we have defined them here. Such programs, including social and natural scientists, are difficult to develop but are useful to explore aspects of fisheries systems generally. Further disciplinary work is required also, for everything from building ocean circulation models to testing how incentive schemes work within fisheries. Research into the design of data collection systems is also needed. The optimum design of a fisheries monitoring program and how it should adapt in response to changes in regulatory regimes is an important issue.

Other research topics should examine what institutional and other changes are needed to ensure greater recognition within governments and industry of the importance of adaptive management approaches and adequate information systems open to public involvement. Some of the consultative mechanisms envisaged within the recent B.C. Strategies for Sustainable Fisheries (despite its overemphasis on the salmon resource alone) may be noted here. More generally, the partnerships envisaged in the amendments to the Fisheries Act tabled in 1996 (but not passed before the 1997 federal election) and a range of other new institutional arrangements offer opportunities for continuous adaptation of organizational structures in response to better information on continuously changing circumstances, although the nature of the necessary accountability mechanisms and institutional measures remains an open question.

5. Informing social decisions through accurate pricing, accounting, reporting, and charging

Principle 5 of Chapter 5 suggests that social and ecological consequences of decisions must be taken into account fully when those decisions are being made. More particularly, it suggests the need to adopt practices of full-cost (ecological) pricing, environmental accounting, and reporting on progress toward sustainability, and to move where appropriate to systems in which the social and ecological costs of resource use and depletion of natural capital are recovered from those causing the impacts (cost recovery). To implement this prescription, a number of steps are essential, as set out below. It must be recognized that these recommendations constitute in some sense a counsel of perfection, unlikely ever to be fully realized. Despite the obvious conceptual, empirical, and logistical difficulties, however, continued effort to move in the directions outlined below is essential if we are to end the ludicrous situation of making decisions on the basis of what can be measured readily using generally accepted accounting conventions rather than what is crucial to the ecosystem. In particular, we must contest the continued error of ignoring the value of natural and social capital in our information and reporting systems, simply because conventional practice has not found ways to assign appropriate numbers to represent that value. The following recommendations suggest alternatives.

Recommendation 5.1. Make markets and price systems work properly to guide decisions in fisheries systems. Prices of goods and services should reflect as far as possible the full opportunity cost of all the resources consumed in their production and use (and the full value of all benefits generated as a result), including social and ecological resources, which may be unpriced or unrecognized, and including all impacts on third parties whether unpriced or not. This recommendation, sometimes expressed as the requirement that "all externalities be internalized" is designed to ensure that price signals work properly to guide production levels and resource use to the point of balance where the value of resources consumed is matched by the value of products or services produced. In other words, for markets to work properly as guides to resource allocation, there must be a complete set of prices properly determined. With our present failure to assign prices to ecological services, natural capital, or social infrastructure, we are far from such a condition.

This "full-cost pricing" argument can be extended to advocacy of "ecological tax reform" designed to ensure that prices are adjusted toward this ideal, precisely because the most dramatic missing pieces in market mechanisms are now recognized to be associated with the failure to price ecological services and natural capital. The magnitude of this omission is suggested by research (Costanza et al. 1997), which estimates the value of unpriced ecological services alone at two to three times the size of reported world gross product. Growing recognition of the importance of social cohesion and cultural solidarity is also leading to an emphasis on attempts to bring into calculations for decision purposes full consideration of social capital and cultural capital, and for that matter human and intellectual capital as well, which are in principle somewhat more amenable to conventional market calculation (Hawken 1993; von Weizsacker 1994).

Recommendation 5.2. Ensure that national accounting and national income measures reflect depletion of marine resources. Price mechanisms provide information as inputs into individual decisions. To report properly, for an enterprise or a society, the consequences of the aggregate of these decisions over a given period, it is necessary to undertake a full accounting of the gains and losses and the resulting balance of assets and liabilities for that organizational unit (United Nations 1993). An organization deludes itself if it counts all current revenue inflow as income without taking into account the resources used up in achieving that inflow. However, this is what organizations and governments in fisheries systems have done traditionally, i.e., treat harvesting as generating income rather than as a transfer of natural capital into financial returns. It is not important when the resources are vast and the current harvests small by contrast such that current decisions can do little to compromise the underlying value of the resource (natural capital). However, now, when the scale of activity is so great and the capacity to harvest so lethal relative to the resource stock itself that a small margin of error could lead to collapse, then counting as income what is really the consumption of the stock (the natural capital) is a potentially fatal error of accounting. With the easing of fiscal constraints in government spending, it is important that the resources, which have been withdrawn from work on environmental satellite accounts and environmental indicators more generally, be reinstated and increased to the point where there is reasonable confidence that Canadians have accurate reporting of the state of the national economy and the Canadian environment insofar as ecological services and natural capital (and marine resources in particular) are concerned. In this respect, the recent important advances in the Canadian System of Environmental and Resource Accounts (Statistics Canada 1997*b*) and continuing work on Canada's National Environmental Indicator Series

(Environment Canada 1997) are encouraging. In collaboration with DFO, a priority task (backed by increased resources if necessary) should be the acceleration of this work to include the information on marine ecosystems for which their conceptual frameworks make provision in principle, but are not yet there.

Recommendation 5.3. Introduce mandatory reporting on progress toward sustainability in fisheries systems. Social reporting should reflect developments in fisheries systems, ocean and coastal marine resources, and coastal communities. While the arguments for ecological pricing and environmental accounting have now been well-developed (though the necessary measures are as yet little implemented), there is less general recognition of the need for a broader mode of reporting on progress toward sustainability as the natural evolution of the recognized need for social accounting. Nevertheless, the case has been fully developed by the National Roundtable on Environment and Economy in its Report to the Prime Minister on Reporting on Progress Toward Sustainability (Hodge et al. 1995), which also urged the need for additional resources for Statistics Canada to assume additional coordinating responsibilities for the purpose.

Recommendation 5.4. Where feasible and appropriate, improve incentives for conservation of marine resources and the social infrastructure (social capital) in coastal communities through user-pay or cost-recovery mechanisms. As noted above, price signals and other data provide information as inputs into decisions of individuals, corporations or communities, and other bodies. To alter incentives, however, or create motivation to act on the basis of the abstract signals provided by these information flows, it may be necessary to reflect this information in charges actually paid or revenues actually received on the basis of the prices quoted. For this reason, the case for full-cost (or ecological) pricing set out in Recommendation 5.1 may be extended to an argument for (full or partial) cost-recovery or user-pay arrangements. When the calculated full-cost price is actually charged (through a cost-recovery system, for example), the users of a natural resource (or of government management services) must consider whether the good or service is truly of sufficient value to them to warrant the costs involved. On the other hand, when the cost charged for use of natural resources does not reflect all the social and ecological costs incurred and the non-use or existence value of the resource to the public owner, then resource exploitation will be excessive and the resources whose services are unpriced in its production are themselves overused. Such ideas have been extensively discussed since at least 1973, when the OECD announced its then-revolutionary "polluter-pay principle," and the justification has been fully articulated in the economics literature for decades. Unfortunately, social practice has not yet caught up with these precepts.

Thus, forcing payment for goods and services on a full-cost basis forces accurate balancing of the value of those goods or services against the true value of the resources used up in generating them. When we are dealing with resources only recently recognized to be scarce and valuable, such as fall under the heading of natural and social capital, but are not yet properly priced or valued, the significance of these measures becomes clear. Without them, we risk making exactly the kinds of ultimately inexcusable mistakes, with serious and possibly irreversible consequences, in the management of these public resources as have resulted in our history of decisions decimating fish stocks and destroying fish habitat.

Examples

Steps toward environmental accounting at the industry and national level are illustrated by the growing popularity of corporate annual environmental reports as supplements to the usual annual reports and by Statistics Canada's promising work on Econnections (Statistics Canada1997*a*), respectively.

Examples which begin to move from environmental accounting to more inclusive reporting on ecosystem health are still in the early stages. Environment Canada's State of Environment Directorate has articulated a comprehensive framework; however, the pace of the work has suffered dramatically in a climate of short-sighted fiscal restraint and there is nothing as yet on marine ecosystems in the "Ecological Life Support Systems" category of the Environmental Indicators Series. A draft report by B.C.'s Commission on Resources and Environment (now disbanded), which was given limited circulation by the Ministry of Environment Lands and Parks in B.C. (Hodge 1997), offers an example of still more general social reporting on progress toward sustainability. Together with the path-breaking Annual Reports of the Provincial Officer of Health in B.C., it illustrates the potential value of such information on long-term social and ecological developments as one basis for assessment of government performance. Again the resources for such work are limited at both federal and provincial levels.

In principle, Multiple Accounts Analysis creates a framework assuring that the social and ecological consequences of proposed action will at least be brought explicitly, if not quantitatively, into decision-making processes. An example of this approach is the multiple-accounts system used by DFO since the late-1970s to evaluate various options for projects in their Salmonid Enhancement Program (SEP). In the early 1990s, this SEP Program changed to a nine-account system (from the five-account system in use since the 1970s) for evaluating potential enhancement projects (activities designed to increase abundance of Pacific salmon, such as habitat restoration, spawning channels, etc.). Each proposed project is now evaluated with respect to potential (*a*) economic benefits, (*b*) employment, (*c*) benefits to First Nations, (*d*) regional development, (*e*) fish manageability, (*f*) fish harvest, (*g*) environment and conservation, (*h*) public involvement, and (*i*) efficiency and effectiveness. In practice, however, some of these accounts (e.g., fish manageability, or the ability to harvest enhanced stocks separately from wild) were given significantly less weight than others, which led to problems in evaluation; debates over the effectiveness of the SEP program have not been eliminated.

Priorities for research and further work

One specific strategy would be a coordinated government initiative to rebuild the monitoring, information, and reporting systems seriously run down or dismantled through the recent period of fiscal restraint. It is interesting to note that in a recent annual report of the Clerk of the Privy Council on the Public Service of Canada, one core role identified for the federal government was the maintenance of national institutions such as Statistics Canada, the Canadian Hydrographic Service, Atmospheric Environment Service, and others, which provide a knowledge base for the entire country. Maintenance of this responsibility demands much greater social investment in efforts to achieve an integrated monitoring of ecosystem integrity and well-being. The example of Statistics Canada coordination of federal and provincial information activities in the fields of population health or justice may suggest models by which federal–provincial cooperation might proceed under the auspices of councils of environment, forests, fisheries, or energy ministers acting jointly for this purpose.

Another strategy would be to upgrade support for research and networking, especially for scientists and managers working on new and emerging fisheries or unanticipated problems possibly over the horizon. In such work, it will be necessary to recognize more fully the value of sustained research and traditional ecological knowledge and to support more integrated work in reporting on progress toward sustainability. For example, see the recently issued Bellagio principles for assessing progress toward sustainability, which suggest the basis for a long-term work program for integrating international reporting efforts (Hardy and Zdan 1997).

6. Implementing institutional reform and promoting commitment to sustainable fisheries

Principle 6 in Chapter 5 suggested that the core task in the renewal of institutions in fisheries systems is to ensure that individual and group roles, rights, and responsibilities are clarified and clearly understood. Institutions and incentive systems must be structured so that personal interests and motivations are aligned as far as possible with group or public interests, especially in sustainability, stewardship, equity, economic efficiency (broadly understood). The discussion there suggested that many different institutional designs might be considered for this purpose, with varying emphasis on market mechanisms, state regulation, or community participation. The strategy we advocate to implement this principle is to assign (or reassign) rights and responsibilities more clearly to reshape incentive systems and get individual incentives right. This approach does not require us as a group to express views about the use of specific management tools or governing instruments, or to try to prescribe which assignment of rights might be appropriate in particular circumstances. However, it does require recognition of the broad spectrum of market, community, and inter-institutional options available.

Below we illustrate some underlying ideas by sketching three apparently successful examples of this strategy in action. The first example is a non-market limited-entry system administered by a joint co-management committee including representatives of government and fishermen; the second is a market-based individual transferable quota (ITQ) system with the Total Allowable Catch (TAC) and seasonal closures administered directly by DFO; and the third is a community quota system managed by community boards. These are only a sample of the rich body of well-documented models embodying various design options (Ostrom 1990; Mitchell 1997; OECD 1997). Our purpose here is simply to illustrate something of the range of options produced by different mixtures and to suggest some areas of agreement on the risks and advantages of various options.

In all three cases, we argue that apparent success lies in the fact that, within an overall allowable catch established by the federal government to assure the primacy of the conservation objective, the assignment of rights and duties effectively decentralizes responsibility for the management and monitoring of detailed harvest and investment plans and decisions. Fishermen, and in some cases coastal communities collectively, all with direct experience and a direct stake in the outcomes, carry the responsibility, individually or as part of a group, to make the necessary judgments, formulate plans, and monitor the results in a changing environment. Thus, each of these schemes, confronted with resource systems displaying different characteristics, emphasizes the link between decisions on the harvesting of uncertain stocks and the consequences for the future availability of fish, and thus moves towards getting the individual incentives right for purposes of sustainability.

It is perhaps useful to emphasize the distinction between management instruments and institutional designs. The Expert Group organized by the OECD Fisheries Committee has reviewed (1997) a range of instruments in use for fisheries management around the world, which it has classified as input controls (licensing, effort limits, etc.), output controls (TAC, individual quotas, community quotas, etc.), or technical instruments (limits on gear, time, or area closures, etc.). Their study predicts the consequences likely to follow from the use of each and attempts to confront these predictions with empirical evidence from observed fisheries. It concludes that the well-known problems associated with open access mean that using TAC as an instrument alone will lead to mismanagement and unsustainable harvesting. From the economic perspective of the OECD Expert Group, the preferred direction of reform is clear: "The results suggest that in order to alleviate these problems [associated with traditional management systems] it would be necessary to introduce right based systems (e.g., transferable individual licenses, individual quotas, and exclusive area use rights)." (OECD 1997, p. 20).

Our work, however, has suggested that to be meaningful one has to take this discussion further, examining who exercises what rights, i.e., who must make which decisions, within the complex nested systems governing human activities in fisheries.

The range of rights and responsibilities involved in managing fisheries and the range of design options possible as these powers are assigned to different participants are illustrated in Table 6.1.

Bundles of rights can be assembled in various ways from this array of component parts. Negotiating public–private partnerships or co-management arrangements will entail coming to agreement on who will exercise which rights and responsibilities and how disputes are to be resolved in the face of continuing change.

It is important to recognize that such bundles of rights are socially constructed (Barzel 1997; Bromley 1991; Sened 1997). Indeed, in all cases, socially enforceable individual rights of access, participation in management, and claims to harvest must be clearly defined and the accompanying codes of conduct must be clearly understood and broadly accepted. The nature of the continuing social constraints on such rights, and the fact that they cannot be absolute or irreversible, must be fully understood. Rules for sharing access to the resource and for sharing the consequences of surprise and shocks in either the natural system or the harvesting and distribution system must be clear. The risks of such unforeseeable change in an indeterminate world should not be considered as fully insured by the taxpayer or as a basis for claims for compensation (Knetsch 1983; Cohen and Radnoff 1998).

In particular, we make the following specific recommendations:

Recommendation 6.1. Clarify the roles, rights, and responsibilities of individuals and groups.

Recommendation 6.2. Design institutions and incentive systems that align individual interests with the goals of biological sustainability, social stability, and economic viability.

Recommendation 6.3. Enhance sense of ownership and secure claim to the resource-market approach and (or) enchance participation by users in decisions on the management and stewardship of the resource-participatory approach.

Examples of alternative institutional arrangements using different instruments

Limited entry in the South Australian abalone fishery. To reduce excess capacity, limited entry was introduced in the South Australian abalone fishery in 1968, when the 200 divers engaged in the industry were granted limited-term licenses of 11-years duration. Sale of licenses was not permitted, on the reasoning that, once sale is permitted, licenses can only be removed through government buy-back. Twelve years after this program was introduced, over half the divers had left the industry voluntarily and their licenses had been retired.

The details of implementation were developed out of intense government–industry debate in the early 1970s and reflected the willingness of government to work with the

Table 6.1. Types of rights held by individuals (I), communities (C), and governments (G) in three management systems.

	Institutional arrangement				
	Limited entry				
Rights and Responsibilities	Non-transferable	Transfer-able	ITQ	Co-management	Community quota
1. Access (to fishing grounds)	I	I	I	C	C
2. Withdrawal	I	I	I	I	C
general					
specific					
3. Alienate access of withdrawal rights	G/C	I	I	C*	C*
4. Limit membership (or quotas or licenses)	C/G	G	G	C/G	C/G
5. Plan harvest					
where	G	G	I	C/G	C/G
when	G	G	I	C/G	C
how	G	G	G	C	C
6. Enforce harvest plan	G	G	I	C/G	C/G
7. Monitor others' harvest					
formal (dockside)	G	G	C (contract)	C/G	C (contract)
informal	n/a	n/a	(Minor)	C	C
8. Cover management costs	G (license fee)	G (license fee)	I (partial)	I/C	I/C
9. Protect/restore habitat	G	G	G	C/G	C/G
10. Enforce habitat protection	G	G	G	C/G	C/G
11. Set policy goals	G	G	G	C/G	C/G
Implement policy	C	G	C	C/G	C/G
12. Research/data collection or analysis	G/C	G	G (or C by contract)	C/G	C/G
13. Coordinate uses/ multiple commons	G	G	G	C/G	C/G
14. Allocate access and TAC	G/C	G	G	C/G	C/G

Note: Specific rights or duties associated with participation in the systems are identified down the left-hand margin. Five different illustrative institutional arrangements are identified across the top. Entries in the cells of the table identify which generic participant (individuals, I; communities, C; governments, G) would hold the right or duty (indicated in the row) under the specified institutional arrangement (indicated in the column).

* No external sale; internal reallocation only.

fishermen in a committee where both were equally represented, to find a system that was biologically sustainable, socially equitable, and obtained optimum economic returns to labor and capital. In working out the new system, government and industry "both now see and understand far more of each other's problems and point of view than ever before. The advent of the Committee has not changed the nature of the problems but has turned them into total industry problems and not just managers' problems." (Stanistreet 1982, p. 142). The committee decided to look at the fishery on a zone-by-zone basis and to work slowly, in order to observe the effects of changes created by new stocks coming on the market and changes created by each phase of the program. Eventually the committee worked out what it considered the optimum number of fishermen for each zone, criteria for interzone transfer of licenses, and a point system for ranking potential new entrants into the fishery, based on residence, years in the fishery, kinship ties to active fishermen, and time on the waiting list. However, because fishermen's recruitment and retirement were slow and fishermen's effort varied, the committee knew that younger entrants would likely increase effort. Therefore, they continued to feel that the number of licenses should not be fixed or transferable through market sales, allowing the committee to respond to new developments in the situation at each point. Through committee decisions, the level of fishing effort could be kept fairly close to maximum sustainable yield without undue risk of stock depletion. The committee could decrease what it considered the optimum number of licenses if newer entrants changed this balance. Although it might have been simpler for government (or the committee) to set the *right* number of licenses and allow them to be transferable, this would have created additional problems later on. As it is, the degree of effort can be adjusted to new conditions in the market or in the productive capacity of the resource, while achieving the goals of sustainable harvest, social stability, and economic viability. (Note that this system focused on a single species only.)

The committee system for monitoring the situation constituted a limited type of co-management at the regional level. Fishermen on the committee had informal rights, shared with government, to limit membership (set the number of entrants in a particular time period), to regulate the transfer of membership (both overall and by area), and to interpret and find appropriate means to implement reforms. In order for the committee to function effectively, a high-level educational process was initially required, involving an up-front investment by government. A significant benefit of that investment was the procurement of informal management rights by fishermen, accompanied by the incurring of responsibilities by fishermen to share with government the burden of explaining and legitimizing the importance of measures being taken. Because this made government's job easier (and perhaps cheaper) in the long run, it was also a benefit to government. The second major benefit of the co-management of licensing was that mutual accountability was established, involving transparent accounting, reporting, and information flow. Thus, both the rights of individual fishermen and the responsibilities of committee members were structured so that pursuit of optimum individual benefits did not prevent the system from achieving the collective (social) objective of appropriate levels of fishing effort.

This limited-entry system, in which the number of licenses and the transferability of licenses were mediated by a co-management committee, displayed several key characteristics that made it work. First, it was adaptive and risk averse in the face of change: the number of licenses was reversible (could be lowered or raised) in the face of new information on resource condition. Second, it was accountable to both government and fishermen through the transparency of standards and rules for conservation established by the co-management committee and by reporting to the fishing community at large. Third, standards and rules for license transfer were based on clearly articulated principles of equity as well as conservation, so that, whether or not fishermen agreed with the policy, they clearly understood why and how it was being implemented and could see that the point system of qualification for licenses was being fairly applied to all.

Halibut ITQs in British Columbia. The halibut ITQ system has been widely praised as the most highly successful example of ITQs in Canada, particularly for the greater profitability it has conferred on the first generation of quota-holders. In B.C., the directed longline fishery for halibut became limited entry in 1979. At that time, a halibut license was separated from a salmon license and the "generalists," fishermen who had not fished halibut in the qualifying years (the late 1970s when they directed their effort at the prosperous salmon fishery), were removed from the fishery. In the 1980s, salmon and herring prices collapsed; halibut license prices then escalated; many licenses changed hands; the fishery became increasingly costly and competitive; and every licensee had to fish hard every year to pay costs. The old system of spreading fishing effort over time was abandoned. Competitive conditions worsened in the fishery until finally DFO was successful in introducing an ITQ system in 1991, following two industry rejections of ITQs in the 1980s.

Under this ITQ system, the TAC and seasonal conservation closures are still set by government but shares of the TAC were allocated to all 435 limited-entry license-holders. Although not all halibut skippers adopted the same strategies, crew size was reduced on average and the total number of trips was spread out over a longer season instead of being compressed into 2–4 trips, the norm before 1991. This change permitted more immediate response to markets and an increase in the quality of the product delivered.

By 1997, only half of the original 435 boats were actively fishing, because fishermen were allowed to transfer 50% of the quota to other fishermen or quota-holders. This means that on average twice the amount of halibut is being landed per boat, resulting in operational economies of scale for the remaining skippers.

Fishermen now pay a set fee per pound landed ($.075 in 1997), mainly to cover the cost of hiring a dockside monitoring company, hiring four additional halibut fishery officers to enforce the program because the fishery is now open longer, and covering all of DFO's direct management expenses for halibut. Fishermen notify the monitors of their intent to fish (hail out) and probable landing date (hail in) at one of 33 specified ports; they must hail in 24 h in advance. Fish are weighed, tagged, and recorded in a central data bank on delivery, and holds are inspected to insure that all fish have been delivered. This fee also pays for at least three meetings per year of the Halibut Advisory Board, made up of fishermen who design the operational rules of the fishery, including landing locations, monitoring, and transfer. Where biological indicators of success are concerned, the evidence suggests that the TAC is being realized and the fishery is being sustainably exploited.

Co-management in the inshore fixed-gear groundfishery of the Scotian Shelf and the Bay of Fundy. The co-management system for fixed-gear fishermen in the Scotia-Fundy region of Atlantic Canada arose originally as a way of avoiding something perceived as worse, namely, the problems associated with transferable individual quotas. ITQs were introduced in the Bay of Fundy herring purse seine fishery in 1982 and in the Nova Scotia crab, shrimp, and midshore groundfish dragger fleet by 1991. By the mid-1990s, the limited-entry fixed-gear inshore groundfishermen, who took cod, haddock, and pollock on set long line and set gillnets, and who had to deal with overcapacity, feared their fishery was headed in the same direction, as government withdrew more and more from former micromanagement roles. These fears led to the establishment of seven regional community-based management boards in 1996.

The establishment of these boards followed a successful pilot community-based management plan in 1995 for one region. Regional plans in five more areas for the 1996 season were developed after region-wide protests in March 1996 over the perceived movement towards ITQs, a substantial increase in licensing fees, and restrictive "core" licensing criteria (which, by itself, would have differentially disadvantaged small-boat fishermen, eventually eliminating 35% of small-boat operators in the Scotia-Fundy fisheries). In October 1996, the program was expanded to seven regional boards on a three-year trial basis.

Each management board fishing in the 4X statistical region has been allocated a share of the total area 4X quotas for cod, haddock, and pollock. Allocations were based on the historical catch of fishermen from each region during the years 1986–1993. For equity considerations, 3% of the quotas from all regions was taken to redistribute to less-advantaged boards to ensure their viability. Thus, the TAC is divided among the seven boards before the season, and according to the management plan of each board, fishermen can fish the entire area, subject to seasons, gear restrictions, weekly trip limits, and by-catch limitations. Adjustments to the local management plans, in response to real-time developments in the fishery, can be made quickly by the board and communicated effectively to participants through a mandatory hail-out/hail-in system. The underlying philosophy of these boards is that the allocations belong to all license-holders in the region and decisions concerning the use of the allocations should be made by a democratic process involving all these license-holders. The workings of the boards permit the development of management plans that are consistent with local fishing patterns, the traditional distribution of catches among gear-types, local ecological conditions, and community needs. Some of the boards are already becoming involved as well in localizing licensing rules, scientific research priorities, and measures for dealing with fleet overcapacity where it exists.

In 1996 management boards also started trading fish with each other. If one board was running out of quota for one species but had an excess of another species, it would make a within-year trade with a board that had an opposite lack/excess of the species in question. Boards are now lobbying DFO to allow uncaught quota to be carried over into the next year, arguing that this would create an incentive to leave fish in the water to grow, rather than forcing the sale or trade of excess quota every year.

At present, these boards have no formal legislative capacity to enforce their management plans. Enforcement has been accomplished instead through contract law, with each board requiring fishermen to sign a contract to follow the management plan and accept penalties for violating the plan as decided by an "infractions committee." The committee can, for example, revoke the dockside monitoring privileges of a fisherman who violates the plan and instruct the monitoring company (hired by the board) to stop monitoring that individual. Because DFO requires monitoring as a condition of licensing, the board's committee can thus essentially revoke or suspend a license. Fishermen refusing to sign a contract or whose license is suspended by a regional board have the option to fish a "generic" management plan as developed by DFO consisting simply of a fishery opening, a limit on the number and the length of fishing trips, and a closing when the overall quota is caught. The fact that the vast majority of fishermen have signed contracts with the management boards attests to their belief that this is their best alternative.

Monitoring and enforcement by the board (through a private monitoring company) is paid for by a $0.01 per pound levy on the catch. DFO enforces the conditions of license, mainly conservation and monitoring requirements, and can impose administrative sanctions for violations. DFO also ensures the integrity of the monitoring programme (accuracy of data, arms-length procedures, and compliance, etc.) and issues daily quota reports from their database to the regional management boards.

The regional boards thus possess five types of informal management rights: (*a*) limited-entry access rights through the collective licenses and catch histories of their members, (*b*) the internal allocation of quota among gear types; (*c*) setting of time, area, and quantity limitations pre-season and in-season; (*d*) restricting of gear in time and space; and (*e*) monitoring and enforcement of in-season regulations. Some boards are already discussing the importance of powers to protect critical habitat and spawning aggregations in their region through measures such as protected areas or time and area closures. DFO is sympathetic to these moves, as evidenced by recent initiatives of their Habitat Division in support of the boards' efforts in this regard. If such measures become institutionalized, the regional boards seem likely to orient themselves toward the management of multispecies and ecosystem linkages.

It is important to note that these rights are restricted by DFO oversight and that the transfer of private access rights (licenses) is somewhat constrained. Although a fisherman may freely sell his limited-entry licenses *within* his board's territory, if he wishes to sell to a party outside this territory, the sale must be negotiated between the two boards.

In summary, the first example above illustrates regulation of *access rights* through limited-entry licensing. This approach is broadly regulatory and could be employed in conjunction with other instruments. In general, it has involved issuing a limited number of fishing permits to those who have the longest and most substantial history in the fishery in question. While many limited-entry systems allow the unencumbered transfer of permits through sale after first issue, this is not necessarily desirable. Here we examined a co-management program based on continually adjusting the number of limited-entry permits and the conditions of transfer (through a joint fishermen–government planning committee). This program has the advantage of increasing economic efficiency (by reducing the number of fishermen) at no increased cost to the remaining fishermen or to government; however, it is not clear how broadly applicable the model may be.

The second example (ITQs) illustrates the use of market-based instruments. It allocates clear *access, withdrawal, and alienation rights* (rights to enter a particular fishery, rights to take a fixed percentage or quota of the harvest, and rights to transfer those rights) to those fishermen who inherit (usually based on catch history) or can acquire (depending on their ability to pay) quota shares. These quasi-property rights provide a secure claim to a share of the resource, and therefore, to a stream of future income. The primary purpose of this instrument is to improve economic performance; by assuring access to a share of the harvest, it reduces the race for fish and the consequent pressures for overcapitalization. It may also create incentives for achieving economies of scale through quota combinations, thereby reducing the fleet and increasing economic efficiency (reducing the capital investment per unit of fish harvested). (From a regional economic perspective, however, reductions in the fleet achieved through such mechanisms may have serious negative impacts on the coastal economy and broader measures of economic performance.) Some observers also argue that secure ownership fosters stewardship by creating incentives to conserve and maximize the value of the resource (McClurg 1997); however, the effectiveness with which it may do so depends on the particular situation and the character of the resource system involved. It has been suggested that ITQ arrangements may increase pressures for dumping and high grading (OECD 1997) and decrease the flexibility of management to utilize in-season adjustments as recommended above. There may be particular concerns attached to the market transferability (alienation) of rights, which introduces speculative motives emphasizing paper gains divorced from the fishery itself, rather than the more fundamental interests of the owner–operator, oriented toward sustainability. Despite these qualifications, the OECD study just cited finds ITQ systems promising instruments in many fisheries. They may also be employed in combination with community-based organizational arrangements,

The third example, based on democratic traditions rather than the market, is community-based co-management (also involving an overall quota, but in this case allocated collectively to a community board). This option allocates *access, withdrawal, and management rights and responsibilities* to a community board, which designs an optimal harvest plan and accompanying regulations that must meet government standards and be open to governmental oversight. Its primary purpose is to improve the design of, and increase compliance with, effective regulations, by having the fishermen (through their representatives on the board) design the plan and regulations. Because fishermen do not have secure individual claims in this case, their well-being depends on the quality of management provided by their own regulations, administered by their own peer group. Stewardship may be enhanced by this position, but whether community commitments can be well-developed and maintained depends on situational factors.

In all three of these examples, the TAC is currently set by the government, although proponents of both market-based and community-based approaches argue that secure ownership or clear understanding of quota rights could lead ultimately to a situation where fisheries management by the effective communities of interest extends to decisions on conservation and investment in future stocks and the enhancement of habitat. However, they differ somewhat in their primary objectives and in their response to the inherent risk in the system and the diversity of individual interests. The market-based ITQ approach leans toward specialization and concentration in the harvesting of a single commercial species. This offers high returns, but, from a stewardship perspective, may be insufficiently risk averse and perhaps not robust in the face of shocks. The community approach is more diversified, spreads the risks over more activities, but may forego, at least in the short term, some margin of the potential current return seen as essential to competitiveness, in favor of resilience in the face of surprise. Also, market-oriented approaches (ITQ systems) do not directly concern themselves with ecological goals, or with social objectives having to do with the distribution of income, the concentration of wealth, or the sustainability of particular communities. Indeed, they also rely on government intervention to limit industrial concentration sufficiently to prevent undue barriers to the entry or survival of new or small competitors in the industry, as well as to assure appropriate redistribution through the capture of resource rents (profits in excess of a normal return on investment) and the taxation of speculative gains achieved through trading of quota.

Research priorities

One research issue of a somewhat instrumental nature is to learn which institutional arrangements can best improve data collection and analysis in both stock and habitat asssessment. For each one, what are the most effective ways for communities or groups of fishermen and governments to work together to ensure a reliable and consistent quality of data, to tap local knowledge of fishermen and communities, and to build more comprehensive regional data bases? What degree of increase in data quality and reliability is possible under conditions where trust and procedures are well-established?

Important research questions to be addressed in the matter of institutional design include effectiveness, reporting, accountability, and financing mechanisms arising in assessing which institutional designs best match the features of particular ecosystems for purposes of aligning individual incentives with social objectives. These questions include the determination of the scale of resource rents that government can collect and the use of these rents. They also include the determination of the balance between public benefits, which ought to be taxpayer-financed as opposed to private benefits, which could legitimately be subject to cost-recovery provisions. As well, the technical problems of accounting for the use of revenues from such cost recovery (to finance fisheries management activities) must be addressed. There is, in addition, the challenge of sharing responsibility and accountability between the Minister, officials, and stakeholders in a "user-pay/user-say" or co-management situation. Such issues of reporting, accountability, power-sharing, and financing obviously must be addressed in detail in assessing alternative institutional designs. For purposes of this report, perhaps the central open question on which work continues to be essential is the extent to which community accountability structures may promote overall goals of sustainability more effectively than market mechanisms involving individual rights. In particular, the effect of limiting or extending transferability, whether of individual or community-held quota, is crucial.

The defining characteristic of an ecosystem approach lies in recognition of the central importance of diversity and complexity. Emphasis is on an interconnected web, networks of nested complex systems, in which the structure as a whole generates many flows and services, not just a single product or species to harvest. In such a multispecies, multiservices setting, defining interests or tenures in harvest of a single species, seem likely to distort incentives and lead to unsustainable behavior jeopardizing the integrity of the system as a whole. The value of the many services not commercially marketed, or of resources not the subject of transferable claims in commercial trade, seems likely to be ignored. It is for this reason that multispecies fisheries are widely recognized to pose problems for ITQ systems.

On the other hand, while community-based co-management systems may succeed in integrating many elements of interconnected ecosystems on a local or regional basis, they face major difficulties in defining the relevant group of stakeholders broadly enough to reflect all legitimate interests at various spatial and temporal scales. Thus, the two outstanding research questions are (*a*) under what conditions individual rights can be responsibly defined in an ecosystem context and (*b*) whether community decisions in ecosystem-based management can be effectively linked to objectives reflecting the interests of all Canadians, who collectively hold the right to authorize access to marine resources in Canadian waters and seek the maximum return from exercising that right.

Conclusion

This chapter has suggested a number of concrete strategies and analytical methods for implementing a precautionary, ecosystem-based approach to decisions in fisheries systems. It has reviewed strategies for improving the information systems on which fisheries management decisions depend. It has addressed the need for major reform of the more general pricing, accounting, and reporting systems, which inform the individual and societal decisions about human harvesting and investment activities on which the sustainability of ecosystems hinges. Finally, the chapter has highlighted the need for conscious design and reform of the social institutions which shape those decisions.

A key theme emerging from the work of this panel, as earlier emphasized, is the necessity to ensure that roles, rights, and responsibilities of all those involved in a complex, changing, and inherently uncertain fishery are clear and clearly understood and that they serve to align private interests with the overall social goal of sustainability. These rights must be interpreted against the difficulties of understanding relationships and anticipating outcomes in this context of continuing change. The commitment to sustainable use of resources may demand changing constraints on the exercise of rights. The commitment to equitable sharing of returns may demand adjustment of roles. In a changing and uncertain world, no one can reasonably claim the right to a guaranteed future revenue stream. All rights, including those termed property, use, or management rights relating to natural capital, are in some sense attenuated, subject to overall social responsibilities for stewardship. Adjustment of such rights must be expected in response to surprises in the system and such adjustments do not create a case for financial compensation. What may be claimed is the right to a fair procedure for adjustment of returns or amendment of constraints on the exercise of rights. If certainty of outcome is not possible, then confidence that fair procedures will prevail in sharing the consequences of unanticipated outcomes is essential.

The array of strategies described in this chapter entails fundamental change in management processes, economic mechanisms, and institutional structures. Above all, it rests on profound change in social attitudes and belief systems. The completion of such a transformation can only be seen as a long-term objective. However, important concrete steps can be taken now. Indeed, a start down these roads must be made now if there is to be any hope that the goal of sustainable fisheries is not to vanish altogether in the near future.

Chapter 7. Summary and Conclusion

1. Scope of this report

The purpose of this chapter is to reiterate briefly the central points of our analysis and highlight what we see as key orientations for action needed now.

This report will not solve all the problems of fisheries management in Canada. Indeed, our work is not directed particularly to the concrete tasks of fisheries management as such. Rather our terms of reference directed us to look more generally at the impacts of human activities on marine resources in the context of continuing change: physical, biological, economic, and social.

Given the uncertainties and risks involved in human activities and the potential for irreversible impacts on interdependent marine resources, we were led to a broad ecosystem approach which includes not only the technical processes of fisheries management, but the political processes and social institutions within which the objectives and frameworks for such management are developed.

In earlier chapters, we reviewed the present state of Canadian fisheries. In later chapters, principles and options were proposed for achieving the three key objectives of biological, social, and economic sustainability in fisheries systems, as components of overall ecosystem integrity. A common theme linking these interconnected objectives is conservation of the marine resource for sustainable use. Integrating these three stated objectives requires seeing beyond the short-term needs of any one component of the system, i.e., natural, economic, or social, to the longer-term needs of fisheries systems as a whole. Ensuring this long-term productive exploitation of living marine resources (fish and other aquatic species) implies a stewardship role, a responsibility for the maintenance of these natural resources into the future.

The marine ecosystems discussed here include humans and their institutional structures. A complex web of dependency now exists as a result of the extensive and growing human exploitation of marine resources. Coastal communities in Canada, indeed worldwide, depend heavily on the marine environment. Our use of ocean resources is now so extensive that the viability of the natural components of the marine ecosystem hinges crucially on successful measures to limit human harvesting activities and habitat destruction.

The emphasis in this report has been on three fundamental characteristics of fisheries systems, i.e., change, complexity, and uncertainty, and on three basic features of human response to them, i.e., conservation, cooperation, and compliance. Our aim was to document and analyze fisheries from a broad, integrated perspective and develop some key principles for guiding those involved in fisheries systems (fishermen, fish processors, residents of fish-dependent communities, scientists, fishery managers, and politicians) as they each attempt to respond to the profound uncertainties surrounding continuing change. Following from these principles are some strategic approaches leading to suggestions for action. We have argued for a positive program of institutional reform reflecting a dramatically altered view of ecosystems and their limits, one that takes into account ongoing change and unavoidable uncertainty. More particularly, we argued for precautionary, ecosystem, integrated, and participatory approaches. The need now is not only to change fisheries management but also to adapt the broader institutions governing fisheries systems in Canada, to respond adequately to ongoing processes of change and to the demands of the resulting social transitions.

This report has illustrated a range of possible institutional designs, in which individuals and economic agents may be led by market forces, cultural traditions, social commitments (implicit or contractual), or negotiated regulations to pursue the goals of biological sustainability, social stability, and economic viability. Strategies for information systems and social structures which support such activities in a general social context were outlined. We also identified more specific strategies or methodologies to guide the decisions of all participants in fisheries systems themselves.

The primary motivation for this report is concern about the present and future state of the fisheries, both in Canada and globally. For ever fewer fisheries is it likely that global catch can be increased substantially on any sustainable basis. However, there is growing demand and growing capacity to catch fish. Within Canada, we have experienced the collapse of the Atlantic groundfisheries, with consequent deep social and economic effects. On both the Pacific and Atlantic coasts and in the Arctic, we have recognized that our fisheries cannot be viewed in isolation or through ideological blinkers. Many fish migrate across national borders and out into the open ocean. Living marine resources are vulnerable to long-range physical and chemical transport systems as well as local discharges and habitat disturbances. Communities that depend on these fish and other marine resources must adopt a broad, pragmatic, and adaptive perspective.

2. Summary: change, complexity, and uncertainty as fundamental characteristics of fisheries systems

Fishery systems are not static. Both fish and humans are affected by constantly changing physical, economic, socio-political, and biological processes. In this report, we have provided examples showing how changes in oceanographic conditions can drive variability in growth and survival rates for fish, cause changes in the distribution of fish, and lead to changes in abundance. Socio-political processes can cause variability in prices and changes in the cost of fishing, the distribution of fishing vessels, and in concepts of property or stewardship and lead to new perspectives on management in an inter-institutional setting.

Linkages within fishery systems illustrate this inherent complexity. Fish depend upon their natural environment; fishing depends upon social and economic conditions as well as the natural environment. Given the complexity and change inherent in this system, it is unreasonable to expect the economic and social yield from fishing effort to be constant. Combined with continuing change, this complexity leads to unavoidable uncertainty. There are linkages between components of fisheries that transmit change and uncertainty. Uncertainty arises from several sources, including climate change, the inherent variability and complexity of fishery systems, imperfect information, errors in observation, errors in implementation, and incomplete compliance with regulations. Therefore, uncertainty about potential effects of climate change on water temperature and currents or survival rates of fish stocks necessarily results in uncertainty in forecasted trends in abundance and productivity of fish populations. No amount of analysis or improved natural and social science will remove such uncertainty, although they may reduce it. Since uncertainty is present and will strongly influence our evaluation of options, this uncertainty must be explicitly included as a component of the information base on which decisions in fisheries will be made.

Given change, complexity, and uncertainty, risk is inevitable for those involved in fisheries. For fish there are risks associated with overexploitation and low abundance. Consequent economic and social risks have become painfully evident in Canada through recent experiences with Atlantic groundfish and Pacific salmon. Uncertainty and risk are always present; both features of fisheries need to be more widely understood and taken into account explicitly by those who make decisions. The present scale of human intervention amplifies the potential risks, with possibly irreversible consequences.

A goal of "conservation leading to sustainable use" integrates the three objectives for Canadian marine fisheries: biological, social, and economic sustainability. The first step in the process for the Canadian public and policy-makers must be development of a conservation ethic to be adopted by those involved in the fisheries. The first half of this report (Chapters 2–4) should provide sufficient information about ongoing processes of change to convince readers of the need for and value of such a conservation ethic. Further, there must be principles and strategies to carry that ethic forward into action and lead to sustainable resource use. Proposed approaches have been discussed in Chapters 5 and 6 and the key ideas are summarized later in this chapter.

It is up to readers of this document to implement these proposed strategies by seeking consensus on measures that might redirect human activities to achieve fisheries that will generate substantial sustainable benefits in the face of ongoing variability, change, and uncertainty.

The conservation ethic we envisage is not merely a set of abstract overarching principles to be spelled out in national and international covenants or codes. As emphasized in Chapter 4, errors in fisheries management creep in at all stages of the process, from estimating the character and state of the system to misjudging the consequences of policy decisions or management action. Ecosystem and precautionary approaches can help in assuring that policy decisions emphasize caution and sustainability in the face of uncertainty in our understanding and knowledge of the system. However, even within a broad ethical framework or general code of conduct, only individual moral codes and commitment to the responsibilities of stewardship can assure that the actions taken from day to day, at sea and ashore, will lean toward sustainable fisheries in the face of uncertainty and unavoidable risk.

Such individual commitment is likely to flow only in a social structure that is seen as legitimate, with management institutions that command confidence. Assurances of voice and participation in overall governing structures must be part of that contract. We are reminded in the summary of section 3 that many of the recommendations contained in this report are directed towards the identification of institutional arrangements and management processes that will command the confidence and promote the mutual respect essential to sustained cooperative management leading to cohesive and sustainable fisheries systems.

3. Summary: conservation, cooperation, and compliance as a basis for managing human activities in fisheries systems

The six core principles set out in Chapter 5 suggest ways in which the goal of sustainable fisheries elaborated in Chapter 1 might be pursued even in the face of the difficult systemic problems identified in Chapters 2–4. These principles, which in Chapter 6 are translated into more concrete strategies, are as follows.

(*a*) Incorporate into decision-making an analysis of structural and dynamic complexities of fishery systems.

(*b*) Incorporate into decision-making an analysis of change, uncertainty, and risk in all fishery activities.

(*c*) Promote and conserve biological, economic, and social diversity.

(*d*) Collect, analyze, and openly communicate essential data and information.

(*e*) Estimate, document, and incorporate into decision-making the social and ecological consequences of decisions and actions.

(*f*) Clearly define roles, rights, and responsibilities of all fishery participants to align their interests with overall objectives of sustainability.

The first two of these principles relate directly to decision processes of individuals and institutions directly involved in fisheries systems in light of the inherent complexity, variability, change, and uncertainty of such systems. Our message is that all participants in management decisions and operational activities must come to recognize that the system and environment in which they work are unavoidably changing and inherently uncertain. Management, investment, harvesting, career, and location decisions cannot be based on an expectation that the future will be like the past or that stocks, which appear to be robust and productive on the basis of recent experience, will remain so in the future. A consciously and indeed aggressively *precautionary approach* (*Strategy 1*) is essential, with recognition that preservation of stocks and biological productivity (and hence employment, incomes, and community survival) in the future will depend on exploitation strategies not pressing too heavily against harvest limits in the present. For this purpose, explicit processes of *risk assessment, risk management, and risk communication* (*Strategy 2*) may often be essential and balanced decision processes incorporating the knowledge and perceptions of all participants in fishery systems will always be central.

Surprises in fishery systems are inevitable. In the face of uncertainty and surprise, a *broad, comprehensive, and continuous flow of information among participants* (*Principle 4 and Strategy 4*) is critical. This information must include announcements or instruction from government managers to fishermen and others in the system as well as the reciprocal flow of fine-grained local knowledge and current monitoring or awareness that comes with intimate involvement in local or traditional activities over years and generations. Also critical is an *integrated approach* (*Principle 3 and Strategy 3*) to the utilization of such information, adaptively, on an ecosystem basis. This integrated approach views the interdependent system as a whole, on spatial and temporal scales appropriate to the multiple stocks and ecological services involved, and with provision for coordination across different jurisdictions involved within the ecosystem as a whole. *Principles 3 and 4 and their associated implementation strategies* thus emphasize the need to promote diversity as a hedge against uncertainty and to work within the framework of an integrated ecosystem-based management process.

More balanced and integrated precautionary decision-making within fisheries systems demands that ecological consequences of proposed actions be identified and assessed before action is taken. Also, to deal more broadly with cross-cutting and cross-sectoral issues, it is necessary to assure that decisions within fisheries systems themselves are integrated within a wider context, including the coastal zone and regional economy as well as the cross-sectoral and cross-departmental "horizontal issues" presently concerning all governments. These larger issues influence the survival of enterprises and communities and call for much more comprehensive and effective procedures for taking into account both social and ecological consequences of management decisions. In the absence of such procedures, serious unintended impacts will be generated. For example, decisions to adopt particular harvesting or processing technologies may lead to excessive but unrecognized costs of by-catch and waste, as well as social dislocation. Indeed, as emphasized in Chapter 6, we consider these questions of by-catch and waste, along with unanticipated or unintended social consequences, to be among the most fundamental challenges in fisheries management.

Methods to achieve better assessment of presently unpriced and unrecognized consequences were emphasized in the discussion of *Principle 5 and Strategy 5*. Similarly, the social consequences for coastal communities and fish-dependent groups, although they may be impossible to price formally, must be taken into account in some fashion in decision processes at all levels in the system. Some of this balancing of concerns undoubtedly occurs now informally in both government and enterprise decisions. Mechanisms to ensure greater integration and coherence of administrative measures in responding to policy dilemmas arising in fisheries is a priority for further research and action.

Several points were made on the questions of incentives, institutional design, enforcement, compliance, and accountability addressed in the discussion of *Principle 6 and Strategy 6*. First, it is worth highlighting that at some level there must be government involvement to represent the interest of the Canadian public in the stewardship of fisheries resources. This government involvement may be simply in establishing general ground rules or more actively in initiating community-based management partnerships, establishing harvest limits, or more specifically, regulating entry, openings, technologies, and practice. However these options are adopted or combined, ultimately there is a need for the general social purpose to be reflected and carried out in fishing effort by individuals exercising their personal discretion, on the basis of socially recognized rights to do so.

In new institutional arrangements, there are many options. The linkage from general strategies and principles to personal conduct might be achieved through contracting by government institutions with agencies ultimately authorizing efforts by individuals or small groups, formal or informal (as envisioned with the provision for "partnerships" in Bill C-62, the new Fisheries Act proposed in 1996). Also, it might occur by delegation to market mechanisms through ITQs to be exercised by individuals who may then group themselves into virtual communities or corporations for purposes of managing the resource more effectively in their collective interest (within the constraints of socially acceptable exercise of these rights). Alternatively, government might allocate quota in a more aggregate form to communities, large or small, formal or informal, that may then choose to organize the involvement of individuals in management plans and harvesting through allocation of individual use rights or through some other mechanisms of community or cooperative (as opposed to corporate) governance.

Experience with some of these institutional arrangements in the contemporary context of a high-technology global economy has not yet led to any final judgments about their relative merits. It is not clear which mechanism mobilizes individual commitment and compliance most effectively, given the ambitious social goals we must pursue in the context of substantial uncertainty about the state of the resource and therefore about fishery systems themselves. From a large and growing literature, no consensus is emerging on even basic questions, such as which approach is more likely to achieve the goal of biological sustainability. There is, however, general agreement that different institutional arrangements are likely to be appropriate in different settings, given the characteristics of particular resource systems. The pragmatic and adaptive approach to institutional innovation mentioned in *Strategy 6* seems the appropriate response under the circumstances.

The initial definition and distribution of access or management rights are crucial and will inevitably be controversial. We have noted earlier that in principle such rights might be auctioned, "grandfathered," or allocated by some other social mechanisms. Also, we see the particular issue of the transferability of rights and the mobility of rights holders (the different impacts and relative merits of non-transferable versus transferable quotas, whether in community or individual form) as a crucial question requiring further research and debate. Because the exercise of rights and subsequent transfers (if permitted) will likely generate excess returns or speculative gains, these should be appropriately taxed. Canadian society, with its collective responsibility for stewardship of the resource, should receive its share of the benefits, while ensuring that appropriate compensation (at market rates) is paid to labor and capital invested in harvesting, processing, and marketing.

Perhaps the most difficult issue is the basic question of who will be excluded and by what means. Who can get past the barriers to entry inevitably created by rationing of access to the common resource and whether this rationing is accomplished by purchasing power, administrative fiat, markets, governments, or communities, remain sensitive questions.

In summary, a number of specific strategies are suggested in Chapter 6 . For implementation of the first two principles, methods to deal with complexity, uncertainty, and surprise were sketched. With respect to the second two principles, the need for integrated ecosystem approaches was emphasized. In dealing with the last two principles, the need for broader consideration of social and ecological consequences, along with more conventional economic and financial concerns to be reflected both in information flows and in institutional arrangements and incentive systems, was stressed. A number of examples were outlined, suggesting that some progress is being made toward implementing these strategies, although few recommendations are being widely applied. We have reviewed these examples (Table 7.1) illustrating the extent to which our recommendations have already been implemented.

Diffusion and adoption of innovative ideas and implementation of corresponding action are slow processes, a fact underlined in Table 7.1. Although many of the ideas behind our six principles are becoming widely accepted and endorsed, action to implement them appears fragmented at best. Some have yet to be accepted as practical and workable. Even where action has been initiated, attainment of the desired outcomes (which reflects the adequacy of the policy design or strategy as well as the extent of implementation effort and the degree of compliance) lies well in the future.

Although much has improved, even in the time since the panel began this report (January 1996), there is still a long way to go before we can claim that the conservation goals enshrined in declarations and the precautionary approaches embedded in policies are truly in place where it counts, in fisheries throughout Canada. Efforts to extend information flow and methods of ecosystem-based management have had only limited success to date, for a variety of reasons including financial constraints and institutional barriers, as well as major conceptual and methodological problems in particular settings. Also, institutional changes to resolve resource conflicts and temper the struggle for shares are still only a dream. Likewise, some of our other of recommendations need further implementation efforts.

We conclude that further action must be taken by all participants in fisheries systems. We must recognize that the long-run synergy and complementarity among the three objectives of sustainable fisheries, linked through the conservation imperative, can be fatally compromised if a focus on short-term goals, i.e., employment, income and revenue generation, or other social or economic factors, press harvesting efforts to levels that leave the biological system critically vulnerable to shocks and surprises that cannot be forecast but are certain to occur.

4. Conclusion

In this report, we argue for an ecosystem approach, risk averse policy, adaptive management, and resilient institutions. Current policy priorities and the policy stance within organizations in fishery systems do seem (as far as statements of intent are to be believed) to be moving toward an appropriate emphasis on the conservation and stewardship goals we espouse. However, as noted, practical progress has been slow and there have been some discouraging steps in the wrong direction.

With some recent measures, DFO and the federal government appear to be embracing the more open, participatory approach that we advocate. (For example, see the new Oceans

Act passed in 1997, the DFO Sustainable Development Strategy discussion paper issued in December 1997, and Richards and Maguire 1998). The search (expressed within the amended Fisheries Act proposed in 1996 as well as in other statements of the Government of Canada) for new institutional arrangements based on devolution of responsibilities and partnership arrangements agrees fully with the clarification of responsibilities within community-based co-management, self-governing structures, or market mechanisms urged here. Furthermore, as noted above, there are promising signs that some members of the Canadian fishing industry are already pursuing some of the principles and strategies outlined here.

The test now is in implementation and in the realization of these ideas and intentions. The challenge will be to create a changed corporate culture and new institutional approaches entrenched in the working structures of government and non-government components of fishery systems. There is a very skeptical audience to be convinced of the reality of the actual commitment to conservation, stewardship, a precautionary approach, meaningful participation, effective co-management, and above all to building and earning trust. This is true from the perspective of both government and industry and there are signs that the war is not won yet. Budget cuts for science and monitoring (both within DFO and Statistics Canada, as well as other agencies and governments) seem extraordinarily shortsighted and in many cases industry and community pressures for maintaining or increasing harvests continue.

As discussed above, we see an opportunity for the federal government, the Department of Fisheries and Oceans, and the fishing industry to move toward more effective partnerships among all who make up the fisheries systems and its institutions (fishermen, cooperatives and corporations, community groups, local and provincial governments, university researchers, and non-government organizations).

In particular, it will be crucial to work out a new role for the federal government and DFO in such a system. The panel notes the need for the Government of Canada to play a clear role in representing the interests of the Canadian public. Also, the core functions for DFO are likely to be more strictly defined than in the past. Perhaps within this redefined role, DFO will serve as guarantor of a more decentralized co-managed system in which the roles, rights, and responsibilities of the individual players are clear and understood, so that they may exercise necessary discretionary operational authority according to their own perceptions and judgments. At the same time, DFO must be open to new approaches and new ideas, responsive to the changing nature of fisheries systems.

Working out the details of such an adjustment to the appropriate core role for DFO would take us beyond the scope of this report. One example of implementing such changes could be in using the concept of precautionary reference points introduced in the 1995 UN Agreement on Straddling Fish Stocks and Highly Migratory Fish Stocks as a conceptual foundation for an appropriate delegation of responsibilities within a broad management network. These reference points specify the overall conservation constraints set by government in the public interest. Given these constraints, responsibilities for adaptive management and the ongoing adjustment of harvest plans in light of changing circumstances and the continuing flow of monitoring information and new knowledge might be carried more fully by the communities and co-management structures closest to the resource and most immediately concerned with the health of the ecosystem. If ongoing monitoring suggests an emerging risk that limit reference points will be violated, pre-agreed measures to counter that risk must come into play to assure that the primacy of the conservation objective is respected.

In pursuing such an approach, early action to meet Canada's long-delayed commitment to ratify UNCLOS and the 1995 Agreement on implementation of its provisions with respect to straddling stocks and migratory stocks obviously is desirable.

Table 7.1. Implementation of strategies proposed in Chapter 6.

Recommendation	Degree of implementation	Examples of progress in implementation
Strategy 1. Implement a precautionary strategy		
1.1. Incorporate analysis of uncertainties into decision-making in management agencies and fishing industries	Partial	Snow crab in the Gulf of St. Lawrence
1.2. Act to reduce uncertainties	Partial	Atlantic striped bass
1.3. Give priority to maintaining productive capacity of fish	Partial	Northwestern Hawaiian Islands lobster
1.4. Maintain critical fish habitat	Partial	Salmon habitat protection in B.C. and marine protected areas
1.5. Keep harvesting capacity below levels sustainable by fish productivity and institute effort controls if capacity exceeds sustainable levels	Partial	Limited entry schemes; effort control in the East Coast lobster fisheries; individual quotas in groundfisheries
1.6. Require comprehensive management plans before onset of fishing	Partial	Post-1983 Norwegian spring-spawning herring King, Tanner, and snow crabs in the Bering Sea
Strategy 2. Implement risk-assessment and risk-management procedures		
2.1. Implement formal procedures for estimating risks	Partial	International Whaling Commission (Kirkwood and Smith 1996)
2.2. Implement formal procedures for risk management that explicitily take into account the results from formal risk assessments	Partial	South African anchovy fishery Setting TACs for cod and haddock stocks on Georges Bank Profit-sharing among B.C. herring boats
2.3. Work cooperatively in project teams to estimate risks	Partial	Fisheries Resource Conservation Council (FRCC) B.C.–Canada MOU for West Coast fishery
2.4. Reverse the burden of proof to allow scientists and managers to use risk-neutral or risk-averse policies consistent with the precautionary approach	Partial	Convention on the Conservation of Antarctic Marine Living Resources (CCAMLR)

Table 7.1 (*continued*).

Recommendation	Degree of implementation	Examples of progress in implementation
	Strategy 3. Promote diversity	
3.1. Recognize the spatial and temporal heterogeneity of fishery systems by adopting an ecosystem-based approach	Partial	Central Region Board's (West Coast of Vancouver Island) integrated approach to terrestrial and marine resorce use
3.3. Create marine protected areas as refugia for exploited species	Partial	Recent federal and B.C. initiatives Recent legal provisions: Oceans Act, new Fisheries Act, Endangered Species Act
Strategy 4. Collect, analyze, and openly communicate essential data and information		
4.1. Implement an integrated, balanced, and transparent information-flow strategy	Partial	Ecological Monitoring and Assessment Network (EMAN)
4.2. Involve fishermen and other groups in the collection of information	Partial	Sentinel Surveys Program on the east coast
4.3. Make better use of existing information resources including traditional knowledge and knowledge of local fishermen	Partial	Global Ocean Ecosystem Program (GLOBEC) Joint management boards under Nunavut and Inuvialuit Final Agreements
4.4. Involve those who collect information in the decision-making process	Partial	Fisheries Resource Conservation Council (FRCC)
4.5. Improve information flow so that it is open and accessible to everyone involved in fishery systems	Partial	FRCC and PFRCC proposal
Strategy 5. Assure informed social decision through pricing, accounting, costing, and reporting		
5.1. Implement full-cost pricing of goods and services to reflect the full cost of all the resources (social and ecological) consumed in their production	None	
5.2. Ensure full accounting of the benefits and costs of fishery systems	Partial	Multiple Accounts Analysis in B.C. Salmonid Enhancement Program's (DFO) multiple account system

Table 7.1 (*concluded*).

Recommendation	Degree of implementation	Examples of progress in implementation
	Strategy 5 (*concluded*).	
5.3. Use a broader mode of reporting on progress toward sustainability	None	BC Commission on Resources and Environment Report NRTEE Report to the Prime Minister
5.4. Move toward fee structures and cost recovery provisions that reflect resource usage	Partial	
Strategy 6. Implement institutional reform and promoting compliance		
6.1. Clarify roles, rights, and responsibilities of individuals and groups	Partial	
6.2. Design institutions and incentive systems that align individual interests with the goals of biological sustainability, social stability, and economic viability	Partial	
6.3. "Traditional" market approaches: enhance sense of ownership and secure claim to the resource	Partial	
Two subclasses:		
(*a*) quota allocated by government (top-down)		(*a*) Halibut ITQ's in British Columbia
(*b*) quota arrangements proposed by fishermen's groups (bottom-up)		(*b*) Geoduck clam fishery in British Columbia
Participatory approaches: enhance participation by users in decisions on the management and stewardship of the resource	Partial	
Two subclasses:		
(*c*) participation through existing regulatory structures		(*c*) Landed entry in the South Australian abalone fishery
(*d*) participation through alternative institutional structures		(*d*) Co-management in the inshore fixed gear Scotia-Fundy 4X groundfishery

Note: Under "degree of implementation," the categories represent the degree to which the recommendation is being implemented in various fisheries in Canada. "None" means no cases; "partial" means some cases exist where the recommendation is being followed; and "extensive" means that the recommendation is widely used. It should be noted that the degree of implementation may have several dimensions: long duration, substantial spatial extent, or perceived degree of success. Of course, most examples are so recent that no significant evaluation of effectiveness is yet possible; therefore, the illustrations cited are largely evidence of intention or feasibility rather than outcome.

The fundamental problem could be viewed essentially as one of equitably regulating access and guiding the allocation of increasingly scarce and valuable common resources, whether at the level of the human species in competition with other species for food and habitat or at the level of the individual fisherman bound by social pressures and personal moral codes. Whether viewed locally or globally, it is interesting to note that this view brings the task back to the basic biological problem outlined in Chapter 2, to manage the competition for fisheries resources to meet social and economic goals while ensuring that the current harvest does not exceed the biological productivity of the stock itself.

In this interpretation, fisheries institutions must mediate between the uncertainty of complex natural systems characterized by ceaseless change and the unpredictability of complex human systems characterized by conflicting interests and distributional concerns. The management of human activities impinging on marine ecosystems will never be exact or perfect, never be purely an economic concern, and never be a matter simply of calculated intervention in a situation fully understood. We do not have human resources sufficient to manage fisheries precisely nor can we control the outcomes or impacts on the resources following from our individual and collective decisions. However, there is some limited power, the power of information, power to exercise authority (when it is accepted as legitimate) and to shape ideas and intervene in processes of decision to influence human activity and the resulting consequences.

The strategies outlined in this document are intended to suggest ways for those involved in fisheries (governments, industry, individuals, and communities) to respond to the need to make important decisions about human intervention in complex systems with only partial information about uncertain developments and dramatically different perceptions of risks held by many people with conflicting interests in the decision processes and their consequences. Not all the counsel in this document can be put into practice easily, nor will all our recommendations be achieved or necessarily endorsed. Indeed, as already emphasized, there is no single institutional solution to be found (and certainly no single policy stance) to offer the best possible structure for all circumstances or all time. However, the effort must be made to develop adaptive processes and structures more consistent with sustainability.

There is a pressing need for continuing progress in the directions outlined above. We must now work collectively towards the application of the identified strategies. A priceless natural and human heritage is at stake.

References

Adkison, M.D., Peterman, R.M., Lapointe, M.F., Gillis, D.M., and Korman, J. 1996. Alternative models of climatic effects on sockeye salmon, *Oncorhynchus nerka*, productivity in Bristol Bay, Alaska, and the Fraser River, British Columbia. Fish. Oceanogr. **5**(3/4): 137–152.

Alverson, D.L. (*Chair*). 1987. A study of trends of cod stocks off Newfoundland and factors influencing their abundance and availability to the inshore fishery. Report of the task group on Newfoundland inshore fisheries. 101 p.

Alverson, D.L., and Larkin, P.A. 1993. Fisheries. Fisheries science and management — Century 21, pp. 150–167. *In* The state of the world's fisheries resource. *Edited by* C.W.Voigtlander. Proceedings of the World's Fisheries Congress. Plenary Session. Oxford and IBH Publishing Co. PVT Ltd., Oxford & New Delhi.

Alverson, D.L., Freeberg, M.H., Murawski, S.A., and Pope, J.G. 1994. A global assessment of fisheries bycatch and discards. Food and Agricultural Organization. Fish. Tech. Pap. **339**. 233 p.

Anderson, D.M. 1989. Toxic algal blooms and red tides: a global perspective. *In* Red tides: biology environmental science and toxicology. *Edited by* T. Okaichi, D.M. Anderson and T. Nemoto. Elsevier, NY. pp. 11–16.

Archibald, C.P., Fournier, D.A., and Leaman, B.M. 1983. Reconstruction of stock history and development of rehabilitation strategies for Pacific ocean perch in Queen Charlotte Sound, Canada. N. Am. J. Fish. Manage. **3**: 283–294.

Arrizaga, A., Buzeta, R., and Fierro, W. 1989. The necessity of integrated coastal development. *In* Artisanal fisheries: towards integrated coastal development. *Edited by* A. Arrizaga. International Development Research Centre, Ottawa, Canada.

Arrow, K., Bolin, B., Costanza, R., Dasgupta, P., Folke, C., Holling, C.S., Jansson, B.-O., Levin, S., Mäler, K.-G., Perrings, C., and Dimentel, D. 1995. Economic growth, carrying capacity and the environment. Science (Washington, D.C.) **268**: 520–521. [Reprinted in Ecological Applications **6**(1) 1996 as part of a special forum. pp. 13–32.]

Ayres, Ian, and Braithwaite, John. 1992. Responsive regulation: transcending the de-regulation debate. Oxford University Press, Oxford.

Bakun, A. 1990. Global climate change and intensification of coastal ocean upwelling. Science (Washington, D.C.) **247**: 198–201.

Bakun, A. 1997. Patterns in the ocean: oceanic processes and marine population dynamics. California Sea Grant. 323 p.

Barrett, L.G. 1984. Capital and the state in Atlantic Canada: the structural context of fishery policy between 1939 and 1977. *In* Atlantic fisheries and coastal communities: fisheries decision-making case studies. *Edited by* C. Lamson and A.J. Hanson. Dalhousie Ocean Studies Programme, Halifax, NS. pp. 77–104.

Barzel, Y. Economic analysis of property rights (2nd ed). Cambridge University Press, Cambridge, MA.

Baumgartner, T.R., Soutar, A., and Ferreira-Bartrina, V. 1992. History of Pacific sardine and northern anchovy populations. CalCOFI Rep. **33**: 24–40.

Beamish, R.J., and Boullion, D.R. 1993. Pacific salmon production trends in relation to climate. Can. J. Fish. Aquat. Sci. **50**: 1002–1016.

Bergh, M.O., and Butterworth, D.S. 1987. Towards rational harvesting of the South African anchovy considering survey imprecision and recruitment variability. S. Afr. J. Mar. Sci. **5**: 937–951.

Berkes, F. and Folke, C. 1994. Linking social and ecological systems for resilience and sustainability. The Beijer International Institute for Ecological Economics. The Royal Swedish Academy of Sciences. Discussion paper No. 52. 16 p.

Beverton, R.J.H., and Holt, S.J. 1957. On the dynamics of exploited fish populations. U.K. Min. Agr., Fish. and Food, Fish. Invest. (Ser. No. 2.), London, U.K.

Beverton, R.J.H. 1993. The RIO Convention and rational harvesting of natural fish resources: The Barents Sea experience in context. *In* Proceedings of the Norway/UNEP Expert Conference on Biodiversity. 24–28 May 1993. *Edited by* O.T. Sandlund and P.J. Schei. Trondheim, Norway. pp. 44–63.

Beverton, R.J.H., and Holt, S.J. 1957. On the dynamics of exploited fish populations. U.K. Min. Agr., Fish. and Food, Fish. Invest. (Ser. 2), **19**: 1–533.

Bigler, B.S., Welch, D.W., and Helle, J.H. 1996. A review of size trends among North Pacific salmon (*Oncorhynchus* spp.). Can. J. Fish. Aquat. Sci. **53**: 455–465.

Blore, I. 1996. How useful to decision-makers is contingent valuation of the environment? Public Admin. Dev. **16**: 215–232.

Boehlert, G.W. 1996. Biodiversity and the sustainability of marine fisheries. Oceanography **9**: 28–35.

Boghen, A.D. (*Editor*). 1989. Cold-water aquaculture in Atlantic Canada. Canadian Institute for Research on Regional Development. Moncton, NB. 410 p.

Botsford, L.W., Castilla, J.C., and Peterson, C.H. 1997. The management of marine ecosystems. Science (Washington, D.C.) **277**: 509–515.

Bradford, M.J. 1995. Comparative review of Pacific salmon survival rates. Can. J. Fish. Aquat. Sci. **52**: 1327–1338.

Bradford, M.J., and Cabana, G. 1997. Interannual variation in stage-specific survival rates and the causes of recruitment variation. *In* Early life history and recruitment in fish populations. *Edited by* R.C. Chambers and E.A.Trippel. Chapman and Hall, London. pp. 469–494.

Brander, K. 1994. Patterns of distribution, spawning, and growth in North Atlantic cod: the utility of inter-regional comparison. *In* Cod and climate change. ICES Marine Sci. Symp. **198**: 406–413.

Brander, K.M. 1995. The effect of temperature on growth of Atlantic cod (*Gadus morhua* L.). ICES J. Mar. Sci. **52**: 1–10.

Broecker, W.S., Peteet, D.M., and Rind, D. 1985. Does the ocean-atmosphere system have more than one stable mode of operation? Nature (Lond.) **315**: 21–26.

Brodeur, R.D., and Ware, D.M. 1992. Long-term variability in zooplankton biomass in the subarctic Pacific Ocean. Fish. Oceanogr. **1**: 32–38.

Brodeur, R.D., and Ware, D.M. 1995. Interdecadal variability in distribution and catch rates of epipelagic nekton in the northeast Pacific Ocean. Can. Spec. Publ. Fish. Aquat. Sci. 121. pp. 329–356.

Bromley, D.W. 1991. Environment and economy: property rights and public policy. Blackwell, Cambridge.

Brown, B.E., and Patil, G.P. 1986. Risk analysis in the Georges Bank haddock fishery — a pragmatic example of dealing with uncertainty. N. Am. J. Fish. Manage. **6**: 183–191.

Buch, E., Horsted, S.A., and Hovgard, H. 1994. Fluctuations in the occurrence of cod in Greenland waters and their possible causes. ICES Mar. Sci. Symp., **198**: 158–174.

Cadigan, S.T. 1995. Marine resource exploitation and development: historical antecedents in the debate over technology in the Newfoundland fishery 1815–1855. Presented at Marine Resources and Human Societies in the North Atlantic since 1500. Memorial University, St. John's, Newfoundland.

Caddy, J.F., and Gulland, J.A. 1983. Historical patterns of fish stocks. Marine policy **7**(4): 267–278.

Caddy, J.F., and Mahon, R. 1995. Reference points for fisheries management. FAO Fish. Tech. Pap. **347**. 83 p.

Canada. 1982*a*. Task Force on Atlantic Fisheries. Report. Ottawa, Canada.

Canada. 1982*b*. The Commission on Pacific Fisheries Policy. Final Report. Ottawa, Canada.

Canada. 1995*a*. The Fraser River Sockeye Public Review Board. Report. Ottawa, Canada.

Canada. 1995*b*. Pacific Policy Roundtable, Report to the Minister of Fisheries and Oceans on the Renewal of the Commercial Pacific Salmon Fishery, Ottawa, Canada.

Calista, Donald J. 1994. Policy implementation. *In* Encyclopedia of policy studies. Second Edition. *Edited by* Stuart S. Nagel. Marcel Dekker, New York.

Cashin, R. 1993. Charting a new course: towards the fishery of the future. Task Force on Incomes and Adjustment in the Atlantic Fishery. Communications Directorate, Fisheries and Oceans, Ottawa, Canada. 199 p.

Castonguay, M., Rose, G.A., and Leggett, W.C. 1991. Onshore movements of Atlantic mackerel (*Scomber scombrus*) in the northern Gulf of St. Lawrence: Associations with wind-forced advections of warmed surface waters. Can. J. Fish. Aquat. Sci. **49**: 2232–2241.

Charles, A.T. 1983. Optimal fisheries investment under uncertainty. Can. J. Fish. Aquat. Sci. **40**: 2080–2091.

Charles, A.T. 1995. The Atlantic Canadian groundfishery: roots of a collapse. Dalhousie Law J. **18**: 65–83.

Charles, A.T. 1998. Living with uncertainty in fisheries: analytical methods, management priorities and the Canadian groundfishery experience. Fish. Res. **37**: 37–50.

Chelton, D.B. 1984. Short-term climate variability in the northeast Pacific Ocean. *In* The influence of ocean conditions on the production of salmonids in the North Pacific. *Edited by* W.G. Pearcy. pp. 87–99.

Chouinard, G.A., and Fréchet, A. 1994. Fluctuations in the cod stocks of the Gulf of St. Lawrence. ICES Mar. Sci. Symp. **198**: 121–139.

Clark, C.W. 1985. Bioeconomic modelling and fisheries management. Wiley-Interscience, New York. 291 p.

Clark, C.W. 1990. Mathematical bioeconomics: the optimal management of renewable resources. 2nd edition. Wiley Interscience, New York.

Clark, C.W., and Lamberson, R. 1982. An economic history and analysis of pelagic whaling. Mar. Policy April: 103–120.

Clark, C.W., and Munro, G.R. 1994. Renewable resources as natural capital: the fishery. *In* Investing in natural capital: the ecological economic approach to sustainability. *Edited by* A-M. Jansson, M. Hammer, C. Folke, and R. Costanza. Island Press, Washington. pp. 43–361.

Clemen, R. 1996. Making hard decisions: an introduction to decision analysis. Duxbury Press, Belmont, CA. 664 p.

Cohen, D., and Radnoff, B. 1998. Regulation, takings, compensation and the environment: an economic perspective. *In* The wealth of forests: markets, regulation and sustainable forestry. UBC Press, Vancouver, BC. (In press).

Colbourne, E., deYoung, B., and Rose, G.A. 1997. Environmental analysis of cod (*Gadus morhua*) migration in relation to the seasonal variations on the Northeast Newfoundland Shelf. Can. J. Fish. Aquat. Sci. **54** (Suppl. 1): 149–157.

Copes, P. 1997. Social impacts of fisheries management regimes based on individual quotas. *In* Social implications of quota systems in fisheries. *Edited by* G. Palsson and G. Petursdottir. Tema NORD 1997: 593, Copenhagen. pp. 61–90.

Costanza, R., d'Arge, R., deGroot, R., Farber, S., Grasso, M., Hannon, B., Limburg, K., Naeem, S., Sutton, P., and van den Belt, M. 1997. The value of the world's ecosystem services and natural capital. Nature (Lond.) **387**: 253–260.

Cove, J.J. 1973. Hunters, trappers, and gatherers of the sea: a comparative study of fishing strategies. J. Fish. Res. Board Can. **30**: 249–259.

Covello, V.T., Von Winterfeldt, D., and Slovic, P. 1986. Risk communication: A review of the literature. Risk Abstr. **3**: 171–182.

Crawford, R. 1989. Exploitation of arctic fishes. Can. Manuscr. Rep. Fish. Aquat. Sci. 2002. 43 p.

Cury, P., and Roy, C. 1989. Optimal environmental window and pelagic fish recruitment success in upwelling areas. Can. J. Fish. Aquat. Sci. **46**: 670–680.

Cushing, D.H. 1973. Dependence of recruitment on parent stock. J. Fish. Res. Board Can. **30**: 1965–1971.

Cushing, D.H. 1995. Population production and regulation in the sea. Cambridge University Press, Cambridge, MA.

Daan, N. 1994. Trends in North Atlantic cod stocks: a critical summary. ICES Mar. Sci. Symp. **198**: 269–270.

Dalley, E.L., and Anderson, J.T. 1997. Age-dependent distribution of demersal juvenile Atlantic cod (*Gadus morhua*) in inshore/offshore northeast Newfoundland. Can. J. Fish. Aquat. Sci. **54**(Suppl. 1): 168–176.

Daly, Herman, and Townsend, Kenneth N. (*Editors*). 1993. Valuing the earth: economics, ecology and ethics. MIT Press, Cambridge, MA.

Dansgaard, W., Johnsen, S.J., Clausen, H.B., Dahl-Jensen, D., Gundestrup, N.S., Hammer, C.U., Hvidberg, C.S., Steffensen, J.P., Sveinbjörnsdottir, A.E., Jouzel, J., and Bond. G. 1993. Evidence for climatic variability of past climate from a 250-year ice-core record. Nature (Lond.) **364**: 218–220.

Davis, A. 1991. Dire straits: the dilemmas of a fishery. The case of Digby Neck and the Islands. Institute of Social and Economic Research, St. Johns, Newfoundland. 111 p.

Davis, M.B., Murphy, E.F., Stansbury, D., and Shelton, P.A. 1996. An assessment of the cod stock in NAFO divisions 3NO. NAFO SCR DOC. 96/80. Ser. No. N2755. 31 p.

Dayton, P.K. 1998. Reversal of the burden of proof in fisheries management. Science (Washington, D.C.) **279**: 821–822.

de Leon, Peter. 1992. The democratization of the policy sciences. Public Admin. Rev. **52**, 2 (March/April): 125–129.

Department of Fisheries and Oceans (DFO). 1976. Policy for Canada's commericial fisheries. Fisheries and Marine Service, Department of the Environment. Ottawa, Canada. 70 p.

Department of Fisheries and Oceans (DFO). 1988. Pacific region salmon stock management plan. I Inner south coast and Fraser River. Discussion document 1986. DFO. 555 West Hastings, Vancouver, BC V6B 5G2. 204 p.

Department of Fisheries and Oceans (DFO). 1993. Lobster summary sheets, 1992–93 season. Annual lobster landings. Fisheries and Oceans Canada, Ottawa, Canada. 22 p.

Department of Fisheries and Oceans (DFO). 1994. (has embraced and affirmed that goal in its own mission and operational mandates).

Department of Fisheries and Oceans (DFO). 1995*a*. Annual summary of fish and marine mammal harvest data for the Northwest Territories. Vol. **6**, 1993–1994. xv + 86 p.

Department of Fisheries and Oceans (DFO). 1995*b*. Federal aquaculture development strategy. Communications Directorate, Fisheries and Oceans Canada, Ottawa, Canada.18 p.

Department of Fisheries and Oceans (DFO). 1995*c* (has embraced and affirmed that goal in its own mission and operational mandates).

Department of Fisheries and Oceans (DFO). 1996*a*. Atlantic groundfish management plan. DFO/5167. Communications Directorate, Fisheries and Oceans Canada, Ottawa, Canada. 232 p.

Department of Fisheries and Oceans (DFO). 1996*b*. The fisheries management partnering concept. DFO/5362. Minister of Public Works and Government Services Canada, Ottawa, Canada.

Department of Fisheries and Oceans (DFO). 1996*c*. Fraser River Coho. *Prepared by* Fraser River Action Plan, Fishery Management Group, Vancouver, BC. March. pp. 1–20.

Department of Fisheries and Oceans (DFO). 1996*d* (has embraced and affirmed that goal in its own mission and operational mandates).

Department of Fisheries and Oceans (DFO). 1996*e*. Landed values, fishing capacity trends and economic performance. DFO Maritimes Regional Fisheries Status Report. No 96/2E. 29 p.

Department of Fisheries and Oceans (DFO). 1997*a*. Working together: a renewed relationship between government and the fishing industry. Minister of Public Works and Government Services Canada, Ottawa, Canada.

Department of Fisheries and Oceans (DFO). 1997*b* (has embraced and affirmed that goal in its own mission and operational mandates).

Department of Fisheries and Oceans (DFO). 1997*c*. An approach to the establishment and management of marine protected areas under the Oceans Act: A discussion paper. January 1997. Ottawa, Canada. Internet address http://www.dfompo.gc.ca/communic/DiscPap/index_e.html

Department of Fisheries and Oceans (DFO). 1998*a*. Sustainable development: a framework for action. (Temporary Version, n.d.). http://www.dfo-mpo.gc.ca/sustdev/sust_ e.htm#nine

Department of Fisheries and Oceans (DFO). 1998*b*. Toward Canada's oceans strategy. Discussion paper. 20 p. [Undated].

deYoung, B., and Rose, G.A. 1993. On recruitment and distribution of Atlantic cod (*Gadus morhua*) off Newfoundland. Can. J. Fish. Aquat. Sci. **50**: 2729–2741.

Dickson, R.R., and Brander, K.M. 1993. Effects of changing windfield on cod stocks of the North Atlantic. Fish. Oceanogr. **2**: 124–153.

Dickson, R.R., Kelly, P.M., Colebrook, J.M., Wooster, W.S. and Cushin, D.H. 1988. North winds and production in the eastern North Atlantic. J. Plankton Res. **10**: 151–169.

Dobell, A.R. 1998. Compliance and constraint: economic instruments in context. *In* The wealth of forests. *Edited by* C. Tollefson. UBC Press, Vancouver, BC.

Doern, G.B., and Conway, T. 1994. The greening of Canada: federal institutions and decisions. Univ. Toronto Press. Toronto, ON.

Doubleday, W.G., Atkinson, D.B., and Baird, J. 1997. Comment: Scientific inquiry and fish stock assessment in the Canadian Department of Fisheries and Oceans. Can. J. Fish. Aquat. Sci. **54**: 1422–1426.

Drinkwater, K.F., Harding, G.C., Mann, K.H., and Tanner, N. 1996. Temperature as a possible factor in the increased abundance of American lobster, *Homerus americanus*, during the 1980s and early 1990s. Fish. Oceanogr. **5**:3/4: 176–193.

Dupont, D. 1996. Limited entry fishing programs: theory and Canadian practice. *In* Fisheries and uncertainty: a precautionary approach to resource management. *Edited by* D.V. Gordon and G.R. Munro. Univ. Calgary Press, Calgary, AB. pp. 107–128.

EMAN (Ecological Monitoring and Assessment Network). 1996. EMAN's goals, objectives and deliverables: 1996 declarations. Occas. Pap. Ser., Rep. No. 3. Internet access http://www.cciw.ca/eman temp/reports/publications/gods96/godd.html.

Environment Canada. 1997. Canada's National Environmental Indicator Series. Internet access http://www.ec.gc.ca

Estes, J.A., Smith, S.S., and Palmisano, J.F. 1978. Sea otter predation and community organization in the western Aleutian Islands, Alaska. Ecology **59**: 822–833.

Fanning, L.P., Mohn, R.K., and MacEachern, W.J. 1996. Assessment of 4VsW cod in 1995 with consideration of ecological indicators of stock status. DFO Atl. Fish. Res. Doc. 96/27. 34p.

FAO (Food and Agriculture Organization of the United Nations). 1995*a*. Review of the state of the world fishery resources: marine fisheries. Fisheries and Agriculture Organization, Marine Resources Service, Fishery Resources Division, Fisheries Department. Fisheries Circular No. 884 FIRM/C884 (En). Rome. 105 p.

FAO (Food and Agriculture Organization of the United Nations). 1995*b*. Review of the state of the world fishery resources: aquaculture. FAO Fish. Circ. 886. Rome, FAO. 127 p.

FAO (Food and Agriculture Organization of the United Nations). 1995*c*. Precautionary approach to fisheries. Part 1: Guidelines on the precautionary approach to capture fisheries and species introductions. FAO Fish. Tech. Pap. 350, Pt. 1. 52 p.

FAO (Food and Agriculture Organization of the United Nations). 1995*d*. The Code of Conduct for Responsible Fisheries. Adopted by the Twenty-eighth session of the FAO Conference in October 1995. 41 pp. Available at FAO web site HTTP://www.fao.org/WAICENT/faoinfo/fishery/FIconde.htm.

FAO (Food and Agriculture Organization of the United Nations). 1996. Precautionary approach to fisheries. Part 2: Scientific papers. Prepared for the technical consultations on the precautionary approach to capture fisheries (including species introductions). FAO Fish. Tech. Pap. 350, Pt. 2. 210 p.

FAO (Food and Agriculture Organization of the United Nations). 1998. FAOSTAT fishery statistics database. FAO Internet WebSite http://apps.fao.org/lim500/nph-wrap.pl? Fishery.Primary&Domain=SUA. FAO, Rome.

Ferson, S. and Ginzberg, L.R. 1997. Different methods are needed to propagate ignorance and variability. Rel. Eng. and Syst. Safety **54**: 133–144.

Fischhoff, B., Slovic, P., and Lichtenstein, S. 1978. How safe is safe enough? Policy Sci. **9**: 2.

Fisheries Resource Conservation Council (FRCC) 1994. Conservation: stay the course. Report to the Minister of Fisheries and Oceans on 1995 Conservation Requirements for Atlantic Groundfish. FRCC 94.R4E, Minister of Supply and Services, Ottawa, Canada. 138 p.

Fisheries Resource Conservation Council (FRCC) 1995. A conservation framework for Atlantic lobster. Report to the Minister of Fisheries and Oceans. FRCC95.R.1. November 1995. 2 p.

Fisheries Resource Conservation Council (FRCC). 1997. Georges Bank. 1997 Conservation requirements for Georges Bank groundfish stocks. Report to the Minister of Fisheries and Oceans. Ottawa, ON. FRCC.1997.R.2.

Fogarty, M.J., and Murawski, S.A. 1998. Large-scale disturbance and the structure of marine systems: fishery impacts on Georges Bank. Ecol. Appl. **8**: S6–S22.

Fogarty, M.J., Rosenberg, A.A., and Sissenwine, M.P. 1992. Fisheries risk assessment: sources of uncertainty. A case study of Georges Bank haddock. Env. Sci. Tech. **26**: 440–447.

Fogarty, M.J., Mayo, R.K., O'Brien, L., Serchuk, F.M., and Rosenberg., A.A. 1997. Assessing uncertainty and risk in exploited marine populations. Rel. Eng. Sys. Safety **54**: 183–195.

Folke, C., Hammer, M., Costanza, R., and Jansson, A. 1994. Investing in natural capital — why, what and how? *In* Investing in natural capital: the ecological economics approach to sustainability. *Edited by* A. Jansson, M. Hammer, C. Folke, and R. Costanza. Island Press, Washington, D.C. 504 p.

Francis, R.I.C.C. 1992. Use of risk analysis to assess fishery management strategies: a case study using orange roughy (*Hoplostethus atlanticus*) on the Chatham Rise, Zealand. Can. J. Fish. Aquat. Sci. **49**: 922–930.

Francis, R.C., and Hare, S.R. 1994. Decadal-scale regime shifts in the large marine ecosystems of the north-east Pacific: a case for historical science. Fish. Oceanogr. **3**: 279–291.

Frank, K.T., Perry, R.I., Drinkwater, K.F., and Lear, W.H. 1988. Changes in the fisheries of Atlantic Canada associated with global increases in atmospheric carbon dioxide: a preliminary report. Can. Tech. Rep. Fish. Aquat. Sci. No. 1652. 52 p.

Frank, K.T., Perry, R.I., and Drinkwater, K.F. 1990. Predicted response of northwest Atlantic invertebrate and fish stocks to CO_2-induced climate change. Trans. Am. Fish. Soc. **119**: 353–365.

Frank, K.T., Drinkwater, K.F., and Page, F.H. 1994. Possible causes of recent trends and fluctuations in Scotian Shelf/Gulf of Maine cod stocks. ICES Mar. Sci. Symp. **198**: 110–120.

Fraser, G.A. 1977. Licence limitation in the British Columbia salmon fishery. Fish. Mar. Serv. Tech. Rep. Pac. Reg. PAC/T-77-13. 74 p.

Fraser River Sockeye Public Review Board. 1995. Fraser River sockeye 1994: problems and discrepancies. Canada Communications Group — Publishing. Ottawa, ON K1A 0S9.

Fréchet, A., Gagnon, Y., Schwab, P., D'Amours, D., Dutil, J.D., Lambert, Y., Bourassa, L., and Rollet, C. 1994. Review of the Northern Gulf of St. Lawrence cod stock (3Pn, 4RS) in 1993. DFO Atl. Fish. Res. Doc. **94/29**. 50 p.

Frederick, S., and Peterman, R.M. 1995. Choosing fishery harvest policies: when does uncertainty matter? Can. J. Fish. Aquat. Sci. **52**: 291–306.

Fryer Report. 1996. Tangled lines: restructuring in the Pacific salmon industry. A Federal-Provincial Review of the Mifflin Plan. Crown Publications, Inc., Victoria, B.C. December 1996. 31 p.

Garcia, S.M., and Newton, C. 1997. Current situation, trends, and prospects in world capture fisheries. *In* Global trends: fisheries management. *Edited by* E.L. Pikitch, D.D. Huppert, and M.P. Sissinwine. Bethesda, MD. Am. Fish. Soc. Symp. **20**: 3–27.

Garrod, D.J., and Schumacher, A. 1994. North Atlantic cod: the broad canvas. ICES Mar. Sci. Symp. **198**: 59–76.

Gell-Mann, Murray. 1994. The Quark and the Jaguar. W.H. Freeman and Company, New York.

Gigerenzer, G. 1994. Why the distinction between single-event probabilities and frequencies is relevant for psychology (and vice versa). *In* Subjective probability. *Edited by* G. Wright and P. Ayton. John Wiley & Sons, Inc., New York. pp. 129–162.

Giguere, M., and Frechette, M. 1994. Scallops on the Quebec coast. *In* Status report on invertebrates 1993: crustaceans and molluscs on the Quebec coast and northern shrimp in the estuary and Gulf of St. Lawrence. *Edited by* L. Savard. Can. Man. Rep. Fish. Aquat. Sci. **2257**. pp. 41–70.

Gilbert, D.G. 1997. Towards a new recruitment paradigm for fish stocks. Can. J. Fish. Aquat. Sci. **54**: 969–977.

Gillis, D.M., Pikitch, E.K., and Peterman, R.M. 1995*a*. Dynamic discarding decisions: foraging theory for high-grading in a trawl fishery. Behav. Ecol. **6**: 146–154.

Gillis, D.M., Peterman, R.M., and Pikitch, E.K. 1995*b*. Implication of trip regulations for high-grading: a model of the behavior of fishermen. Can. J. Fish. Aquat. Sci. **52**: 402–415.

Gislason, H. 1994. Ecosystem effects of marine fishing activities in the North Sea. Mar. Pollut. Bull. **29**: 520–527.

Gislason, G. 1996. Fishing for answers: coastal communities and the BC salmon fishery. ARA Consulting Group Inc., Vancouver, BC.

Gordon, H.S. 1954. The economic theory of a common property resource: the fishery. J. Political Econ. **62**: 124–142.

Gordon, D.V., and Munro, G.R. (*Editors*). 1996. Fisheries and uncertainty: a precautionary approach to resource management. University of Calgary Press, Calgary, AB. 195 p.

Gough, J. 1993. A historical sketch of fisheries management in Canada. *In* Perspectives on Canadian marine fisheries management. *Edited by* L.S. Parsons and W.H. Lear. Can. Bull. Fish. Aquat. Sci. **226**. pp. 5–53.

Gregory, R., Brown, T.C., and Knetsch, J.L. 1996. Valuing risks to the environment. Annals, AAPSS **545**: 54–63.

Griswold, B.L. 1997. Fisheries and pollution. Trans. Am. Fish. Soc. **126**: 504–505.

Groot, C., and Margolis, L. (*Editors*). 1991. Pacific salmon life histories. UBC Press, Vancouver, BC. 564 p.

Grosslein, M.D., Langton, R., and Sissenwine, M.P. 1980. Recent fluctuations in pelagic fish stocks in the Northwest Atlantic, Georges Bank region in relation to species interactions. Rapp. P-V Reun. Cons. Int. Explor. Mer **177**: 374–405.

Grumbine, R.E. 1994. What is ecosystem management? Conserv. Biol. **6**: 27–38.

Gubbay, S. (*Editor*). 1995. Marine protected areas. Chapman and Hall, London. 232 p.

Gulland, J.A. (*Editor*). 1977. Fish Population Dynamics. Wiley Interscience, New York.

Gulland, J.A. 1983. Fish stock assessment: a manual of basic methods. John Wiley & Sons, Inc., New York. 223 p.

Gulland, J.A., and Boerema, L.K. 1973. Scientific advice on catch levels. Fish. Bull. (US). **75**: 325–335.

Guppy, N. 1987. Labouring on shore: transforming uncommon property into marketable products. *In* Uncommon property: the fishing and fish processing industries in British Columbia. *Edited by* P. Marchak, N. Guppy, and J. McMullan. UBC Press, Vancouver, BC. pp. 199–222.

Haedrich, R.L. 1995. Perspectives on the Canadian east coast fishery crisis. *In* Getting the missing fish story straight. *Edited by* Patricia Gallaugler. Simon Fraser University, Vancouver, BC. pp. 12–19.

Hanna, S., and Munasinghe, M. (*Editors*). 1995. Property rights and the environment: social and ecological issues. Beijer Institute of Ecological Economics, Stockholm, Sweden and the World Bank, Washington, D.C. 164 p.

Hardy, P., and Zdan, T. 1997. Assessing sustainable development: principles in practice. International Institute for Sustainable Development, Winnipeg, MB.

Hare, G.M., and Dunn, D.L. 1993. A retrospective analysis of the Gulf of St. Lawrence snow crab (*Chionecetes opilio*) fishery, 1965–1990. *In* Perspectives on Canadian marine fisheries management. *Edited by* L.S. Parsons and W.H. Lear. Can. Bull. Fish. Aquat. Sci. 226. pp. 267–293.

Hare, S.R., and Francis, R.C. 1995. Climate change and salmon production in the Northeast Pacific Ocean. *In* Climate change and northern fish populations. *Edited by* R.J. Beamish. Can. Spec. Publ. Fish. Aquat. Sci. 121. pp. 357–372.

Harris, L. (*Chair*). 1990. Independent review of the state of the northern cod stock. Report prepared for the Minister of Fisheries. Communications Directorate, Canada. Department of Fisheries and Oceans, Ottawa, Canada. 155 p. + appendices.

Hawken, Paul. 1993. The ecology of commerce: a declaration of sustainability. Harper Collins, New York.

Helle, J.H., and Hoffman, M.S. 1995. Size decline and older age at maturity of two chum salmon (*Oncorhynchus keta*) stocks in western North America, 1972–1992. *In* Climate change and northern fish populations. *Edited by* R.J. Beamish. Can. Spec. Publ. Fish. Aquat. Sci. 121. pp. 245–260.

Helton, J.C., and Burmaster, D.E. (*Editors*). 1996. Treatment of aleatory and epistemic uncertainty. Special Issue, Reliability Engineering and System Safety, Vol. 52. pp. 91–257.

Hennemuth, R.C., Palmer, J.E., and Brown, B.E. 1980. A statistical description of recruitment in eighteen selected fish stocks. J. Northwest Atl. Fish. Sci. **1**: 101–111.

Hess, Charlotte. 1996. Common pool resources and collective action: a bibliography, Vol. 3. Workshop in Political Theory and Policy Analysis. Indiana University, Bloomington, IN.

Hess, Charlotte. 1998. Common pool resources and collective action: a bibliography, Vol. 4. (In press). Available electronically at: http://www.indiana.edu/~libref/cprbib

Hilborn, R., and Peterman, R.M. 1996. The development of scientific advice with incomplete information in the context of the precautionary approach. *In* Precautionary approach to fisheries. Part 2: Scientific papers. Prepared for the Technical Consultation on the Precautionary Approach to Capture Fisheries (Including species introductions). Rome. FAO Fish. Tech. Pap. 350. pp. 77–101.

Hilborn, R., and Walters, C.J. 1992. Quantitative fish stock assessment. Chapman and Hall, Inc., New York. 570 p.

Hilden, M. 1997. Conflicts between fisheries and seabirds — management options using decision analysis. Mar. Policy **21**(2): 143–153.

Hodge, R.A. 1997. Progress BC: an assessment of British Columbia's progress toward sustainability. *In* Assessing sustainable development: principles in practice. *Edited by* P. Hardy and T. Zdan. International Institute for Sustainable Development, Winnipeg, MB. pp. 37–52.

Hodge, R.A., Holtz, S., Smith, C., and Hawke-Baxter, K. 1995. Pathways to sustainability: assessing our progress. National Roundtable on Environment and Economy, Ottawa, Canada.

Holling, C.S., and Meffe, G.K. 1996. Command and control and the pathology of natural resource management. Conserv. Biol. **10**: 328–337.

Hollowed, A.B., and Wooster, W.S. 1992. Variability of winter ocean conditions and strong year classes of northeast Pacific groundfish. ICES Mar. Sci. Symp. **195**: 433–444.

Hollowed, A.B., and Wooster, W.S. 1995. Decadal-scale variations in the eastern subarctic Pacific: II. Response of northeast Pacific fish stocks. *In* Climate change and northern fish populations. *Edited by* R.J. Beamish. Can. Spec. Publ. Fish. Aquat. Sci. 121. pp. 373–385.

Huppert, D.D. 1996. Risk assessment, economics and precautionary fishery management. *In* Precautionary approach to fisheries. Part 2: Scientific papers. Prepared for the Technical Consultation on the Precautionary Approach to Capture Fisheries (Including Species Introductions). FAO Fisheries Technical Paper No. 350, Part 2. FAO, Rome. pp. 103–127.

Hutchings, J.A. 1996. Spatial and temporal variation in the density of northern cod and a review of hypotheses for the stock's collapse. Can. J. Fish. Aquat. Sci. **53**: 943–962.

Hutchings, J., and Myers, R.A. 1994. What can be learned from the collapse of a natural resource? Atalntic cod, *Gadus morhua*, off Newfoundland and Labrador. Can. J. Fish. Aquat. Sci. **51**: 2126–2146.

Hutchings, J., and Myers, R.A. 1995. The biological collapse of Atlantic cod off Newfoundland and Labrador: an exploration of historical changes in exploitation, harvesting technology, and management. *In* The North Atlantic fishery: strengths, weaknesses and challenges. *Edited by* R. Arnason and E.F. Felt. Inst. Island Stud. Univ. Prince Edward Island, Charlottetown, PEI. pp. 37–93.

Hutchings, J. A., Myers, R.A., and Lilly, G.R. 1993. Geographic variation in the spawning of Atlantic cod, *Gadus morhua*, in the Northwest Atlantic. Can. J. Fish. Aquat. Sci. **50**: 2457–2467.

Hutchings, J.A., Walters, C., and Haedrich, R.L. 1997. Is scientific inquiry incompatible with government information control? Can. J. Fish. Aquat. Sci. **54**: 1198–1210.

ICES. 1993. Reports of the ICES Advisory Committee on Fishery Management 1992. ICES Cooperative Research Report No. 193.

IJC (International Joint Commission). 1995. 1993–95 Priorities and Progress under the Great Lakes Water Quality Agreement. August, 1995.

IPCC (Intergovernmental Panel on Climate Change). 1996. Climate change 1995, the science of climate change. *Edited by* J.T. Houghton, L.G. Meira Filho, B.A. Callander, N. Harris, A. Kattenberg, and K. Maskell. Cambridge University Press, Cambridge, U.K. 572 pp.

IPHC (International Pacific Halibut Commission). 1997. Annual meeting handout, Seattle, Washington. August, 1995.

IPSFC (International Pacific Salmon Fisheries Commission). 1984. IPSFC Annual Report for 1983. IPSFC, New Westminster, BC.

Ishida, Y., Ito, S., Kaeriyama, M., McKinnell, S., and Nagasawa, K. 1993. Recent changes in age and size of chum salmon (*Oncorhynchus keta*) in the North Pacific Ocean and possible causes. Can. J. Fish. Aquat. Sci. **50**: 290–295.

Jasanoff, S. 1992. Pluralism and convergence in international science policy. *In* Science and sustainability. International Institute for Applied Systems Analysis, Vienna.

Jentoft, S., and McCay, B. 1995. User participation in fisheries management. Lessons drawn from international experience. Mar. Policy **19**(3): 227–246.

Johnson, C.R., and Mann ,K.H. 1988. Diversity, patterns of adaptation, and stability of Nova Scotian kelp beds. Ecol. Monogr. **58**(2): 129–154.

Kahneman, D., and Tversky, A. 1979. Prospect theory: an analysis of decision under risk. Econometrica **47**: 263–291.

Kearney, J.F. 1984. The transformation of the Bay of Fundy herring fisheries, 1976–1978: an experiment in fishermen–government co-management. *In* Atlantic fisheries and coastal communities: fisheries decision-making case studies. *Edited by* C. Lamson and A.J Hanson.

Dalhousie Ocean Studies Programme, Halifax, NS. pp. 165–204.

Keeney, R.L. 1982. Decision analysis: an overview. Operations Res. **30**: 803–838.

Kerr, S.R., and Ryder, R.A. 1997. The Laurentian Great Lakes experience: a prognosis for the fisheries of Atlantic Canada. Can. J. Fish. Aquat. Sci. **54**: 1190–1197.

Kendall, A.W. Jr, Perry, R.I., and Kim, S. (*Editors*). 1996. Fisheries oceanography of walleye pollock in Shelikov Strait, Alaska. Fish. Oceanogr. **5** (Suppl. 1). 203 p.

Kirkwood, G.P., and Smith, A.D.M. 1996. Assessing the precautionary nature of fisheries management strategies. *In* Precautionary approach to fisheries. Part 2: Scientific papers. Prepared for the Technical Consultation on the Precautionary Approach to Capture Fisheries (Including Species Introductions). FAO Fish. Tech. Pap. No. 350/2. FAO, Rome. pp. 141–158.

Knetsch, J.L. 1983. Property rights and compensation. Butterworths, Toronto, ON.

Knetsch, J.L. 1993. Environmental valuation: some practical problems of wrong questions and misleading answers. Resource Assessment Commission, Canberra, Australia. Occasional Publication No. 5. pp. 1–26.

Kowalik, Z., and Murty, T.S. 1993. Numerical modelling of ocean dynamics. World Scientific, Singapore. 481 p.

Kruzynski, G.M., and Birtwell, I.K. 1994. A predation bioassay to quantify the ecological significance of sublethal responses of juvenile chinook salmon to the antisapstain fungicide TCMTB. Can. J. Fish. Aquat. Sci. **51**: 1780–1790.

Langdon, S. 1982. Managing modernization: a critique of formalist approaches to the Pacific salmon fisheries. *In* Modernization and marine fisheries policy. *Edited by* John Maiolo and Michael Orbach. University of Michigan Press, Ann Arbor, MI.

Lapointe, M.F., Peterman, R.M., MacCall, A.D. 1989. Trends in fishing mortality rate along with errors in natural mortality rate can cause spurious trends in fish stock abundances estimated by virtual population analysis (VPA). Can. J. Fish. Aquat. Sci. **46**: 2129–2139.

Lasker, R. 1975 Field criteria for survival of anchovy larvae: the relation between inshore chlorophylll maximum layers and successful first feeding. Fish. Bull. (U.S.) **73**: 453–462.

Lawson, P. 1993. Cycles in ocean productivity, trends in habitat quality and restoration of salmon runs in Oregon. Fisheries (Bethesda) **188**: 6–10.

Leaman, B.M. 1993. Reference points for fisheries management: the western Canadian experience. *In* Risk evaluation and biological reference points for fisheries management. *Edited by* S.J. Smith, J.J. Hunt, and D. Rivard. Can. Spec. Publ. Fish Aquat. Sci. 120. pp. 15–30.

Leaman, B.M., and Beamish, R.J. 1984. Ecological and management implications of longevity in some northeastern Pacific groundfishes. Int. North Pac. Fish. Comm. Bull. **42**: 85–97.

Levy, D.A. (*Editor*). 1996. Strait of Georgia fisheries sustainability review. Hatfield Consultants Ltd., West Vancouver, BC. 441 p.

Lichbach, Mark I. 1996. The cooperator's dilemma. Michigan University Press, Ann Arbor, MI.

Lindblom, C.E., and Cohen, D.K. 1979. Usable knowledge: social science and social problem solving. Yale University Press, New Haven, CT.

Linder, E., Patil, G.P., and Vaughan, D.S. 1987. Application of event tree risk analysis to fishery management. Ecol. Model. **36**: 15–28.

Loch, J.S., Moriyasu, M., and Jones, J.B. 1995. An improved link between industry, management, and science: review of case history of the southwestern Gulf of St. Lawrence snow crab fishery. Aquat. Living Resour. **8**: 253–265.

Lynch, D.R., and Davies, A.M. (*Editors*). 1995. Quantitative skill assessment for coastal ocean models. American Geophysical Union, Washington, D.C. 510 p.

Macdonald, R.D.S. 1984. Canadian fisheries policy and the development of Atlantic coast groundfisheries management. *In* Atlantic fisheries and coastal communities: fisheries decision-making case studies. Dalhousie Ocean Studies Programme, Halifax, NS. pp. 15–75.

MacDonald, J.M. 1995. Appreciating the precautionary principle as an ethical evolution in ocean management. Dev. Int. Law **26**: 255–286.

Magnuson, J.J., (*Chair*), Allendorf, F.W., Beschta, R.L., Bisson, P.A., Carson, H.L., Chapman, D.W., Hanna, S.S., Kapuscinski, A.R., Lee,

K.N., Lettenmaier, D.P., McCay, B.J., MacNabb, G.M., Quinn, T.P., Riddell, B.E., and Werner, E.E. 1996. Upstream: salmon and society in the Pacific Northwest. National Academy Press, Washington, D.C. 452 p.

Maguire, L.A., and Boiney, L.G. 1994. Resolving environmental disputes: a framework incorporating decision analysis and dispute resolution techniques. J. Environ. Manage. **42**: 31–38.

Maguire, J.-J., and Mace, P.M. 1993. Biological reference points for Canadian Atlantic gadoid stocks. *In* Risk evaluation and biological reference points for fisheries management. *Edited by* S.J. Smith, J.J. Hunt, and D. Rivard. Can. Spec. Publ. Fish. Aquat. Sci. 120. pp. 321–332.

Malloy, K.D., Holman, M.A., Mitchell, D., and Dietrich, H.W., III. 1997. Solar UVB- induced DNA damage and photoenzymatic DNA repair in antarctic zooplankton. Proc. Nat. Acad. Sci. (USA) **94**: 1258–1263.

Manabe, S., and Stouffer, R.J. 1995. Simulation of abrupt climate change induced by freshwater input to the North Atlantic Ocean. Nature (Lond.) **378**:165–167.

Mangel, M., Talbot, L.M., Meffe, G.K., Agardy, M.T., Alverson, D.L., Barlow, J., Botkin, D.B., Budowski, G., Clark, T., Cooke, J., Crozier, R.H., Dayton, P.K., Elder, D.L., Fowler, C.W., Funtowicz, S., Giske, J., Hofman, R.J., Holt, S.J., Kellert, S.R., Kimball, L.A., Ludwig, D., Magnusson, K., Malayang, B.S., III, Mann, C., Norse, E.A., Northridge, S.P., Perrin, W.F., Perrings, C., Peterman, R.M., Rabb, G.B., Regier, H.A., Reynolds, J.E., III, Sherman, K., Sissenwine, M.P., Smith, T.D., Starfield, A., Taylor, R.J., Tillman, M.F., Toft, C., Twiss, J.R., Wilen, J., and. Young, T.P. 1996. Principles for the conservation of wild living resources. Ecol. Appl. **6**(2): 338–362.

Mann, K.H., and Lazier, J.R.N. 1996. Dynamics of marine ecosystems: biological-physical interactions in the oceans. 2nd ed. Blackwell Science, Cambridge, MA.

March, James G., and Olsen, Johan P. 1989. Rediscovering institutions: the organizational basis of politics. The Free Press, New York.

Marchak, P., Guppy, N., and McMullan, J. (*Editors*). 1987. Uncommon property: the fishing and fish processing industries in British Columbia. UBC Press, Vancouver, BC. 402 p.

McAllister, M.K., and Peterman, R.M. 1992. Experimental design in the management of fisheries: a review. N. Am. J. Fish. Manage. **12**: 1–18.

McAllister, D.E., Parker, B.J., and McKee, P.M. 1985. Rare, endangered and extinct fishes in Canada. Syllogeus **54**. 192 p.

McCartney, M., Curry, R., and Bezdek, H. 1997. North Atlantic's transformation pathway. Oceanus **39**(2): 19–23.

McClurg, T. 1997. Bureaucratic management versus private property: ITQs in New Zealand after ten years. *In* Fish or cut bait! The case for individual transferable quotas in the salmon fishery of British Columbia. *Edited by* L. Jones and M. Walker. The Fraser Institute, Vancouver, B.C.

McKinnell, S. 1995. Age-specific effects of sockeye abundance on adult body size of selected British Columbia sockeye stocks. Can. J. Fish. Aquat. Sci. **52**: 1050–1063.

Mitchell, D.A. 1997. Sustainable by design. Ph.D. dissertation, University of Victoria, Victoria, BC.

Moffett, J. 1996. Environmental priority setting based on comparative risk and public input. Can. Public. Admin. **39**: 362–385.

Morgan, M.G., and Henrion, M. 1990. Uncertainty: a guide to dealing with uncertainty in quantitative risk and policy analysis. Cambridge University Press, Cambridge, MA. 332 p.

Morgan, M.G., Fischhoff, B., Bostrom, A., Lave, L., and Atman, C.J. 1992. Communicating risk to the public. Envir. Sci. Technol. **26**: 2048–2056.

Morgan, M.J., Bishop, C.A., and Baird, J.W. 1994. Temporal and spatial variation in age and length at maturity in cod in Divisions 2J and 3KL. NAFO Sci. Counc. Studies. **21**: 83–89.

Morgan, M.J., DeBlois, E.M., and Rose, G.A. 1997. An observation on the reaction of Atlantic cod (*gadus morhua)* in a spawning shoal to bottom trawling. Can. J. Fish. Aquat. Sci. **54**: 217–223.

Morgenstern, Oskar. 1963. On the accuracy of economic observations, 2nd edition, completely revised. Princeton University Press, Princeton, Princeton, NJ.

Mullin, M.M. 1995. The Californian El Nino of 1992 and the fall of *Calanus*. CalCOFI Rep. **36**: 175–178.

Munro, G., and Neher, P. 1995. Institutional change and the management of British Columbia fisheries. *In* Managing natural resources in British Columbia: markets, regulations, and sustainable development. *Edited by* A. Scott, J. Robinson, and D. Cohen. UBC Press, Vancouver, BC.

Munro, G.R., and Scott, A.D. 1985. The economics of fisheries management. *In* Handbook of natural resource and energy economics. Vol. II. *Edited by* A.V. Kneese, and J.L. Sweeney. North-Holland Publisher, Amsterdam. pp. 623–676.

Murphy, G. 1977. Clupeoids. *In* Population dynamics. *Edited by* J.A. Gulland. John Wiley & Sons Inc., New York. pp. 283–308.

Myers, R.A., Mertz, G., and Bishop, C.A. 1993*a*. Cod spawning in relation to physical and biological cycles of the northern North-West Atlantic. Fish. Oceanogr. **2**(3/4): 154–165.

Myers, R.A., Drinkwater, K.F., Barrowman, N.J., and Baird, J.W. 1993*b*. Salinity and recruitment of Atlantic cod (*Gadus morhua*) in the Newfoundland region. Can. J. Fish. Aquat. Sci. **50**: 1599–1609.

Myers, R.A., Mertz, G., and Barrowman, N. 1995. Spatial scales of variability in cod recruitment in the North Atlantic. Can. J. Fish. Aquat. Sci. **52**: 1849–1862.

Mysak, L.A. 1986. El Nino, interannual variability and fisheries in the northeast Pacific Ocean. Can. J. Fish. Aquat. Sci. **43**: 464–497.

Naeem, S., Thompson, L.J., Lawler, S.P., Lawton, J.H., and Woodfin, R.M. 1994. Declining biodiversity can alter the performance of ecosystems. Nature (Lond.) **368**: 734–737.

National Academy of Public Administration (NAPA). 1995. Setting priorities, getting results: a new direction for EPA. Report to Congress, Washington, D.C. April 1995.

National Academy of Sciences (NAS). 1998. Improving fish stock assessments. Committee on Fish Stock Assessment Methods, National Research Council, Washington, D.C., National Academy of Sciences Press, Washington, D.C. 188 p.

National Oceanic and Atmospheric Administration (NOAA). 1992. Report of the NOAA Panel on Contingent Valuation. Nat. Register **58**(10): 4601–4614.

National Oceanic and Atmospheric Administration (NOAA). 1993. Our living oceans. Report on the Status of U.S. Living Marine Resources 1993. NOAA Tech. Memo NMFS-F/SPO-15. National Oceanic and Atmospheric Administration, Washington, D.C.

National Oceanic and Atmospheric Administration (NOAA). 1996. Our living oceans. Report on the Status of U.S. Living Marine Resources 1995. NOAA Tech. Memo NMFS-F/SPO-19. National Oceanic and Atmospheric Administration. Washington, D.C.

National Research Council (U.S.) (NRC). 1996. The Bering Sea ecosystem. Committee on the Bering Sea Ecosystem, National Research Council. National Academy Press, Washington, D.C. 307 p.

Norgaard, R.B. 1994. Development betrayed: the end of progress and a coevolutionary revisioning of the future. Routledge, London, U.K.

Olver, C.H. Shuter, B.J., and Minns, C.K. 1995. Toward a definition of conservation principles for fisheries management. Can. J. Fish. Aquat. Sci. **52**: 1584–1594.

Organization for Economic Cooperation and Development (OECD). 1981. The welfare state in crisis (especially papers by A. R. Dobell and K. Eide). OECD, Paris.

Organization for Economic Cooperation and Development (OECD) 1997. Towards sustainable fisheries: economic aspects of the management of living marine resources. OECD, Paris.

Ommer, R.E. 1994. One hundred years of fishery crises in Newfoundland. Acaduensus **XXIII**(2): 5–20.

O'Riordan T., and Cameron, J. 1994. Interpreting the precautionary principle. Earthscan, London.

Ostrom, E. 1990. Governing the commons. The evolution of institutions for collective action. Cambridge University, New York. 280 p.

Pacific Salmon Alliance (PSA). 1997. Regional co-management in Pacific salmon fisheries. B.C. Fisheries Secretariat, Victoria, BC. 25 p.

Parsons, L.S. 1993. Management of marine fisheries in Canada. Can. Bull. Fish. Aquat. Sci. 225. 763 p.

Parsons, T.R. 1992. The removal of marine predators by fisheries and the impact of trophic structure. Mar. Pollut. Bull. **25**: 51–53.

Parsons, L.S., and Lear, W.H. (*Editors*). 1993. Perspectives on Canadian marine fisheries managment. Can. Bull. Fish. Aquat. Sci. 226. 446 p.

Pauly, D. 1980. On the interrelationships between natural mortality, growth parameters, and mean environmental temperature in 175 fish stocks. J. Conseil **39**(2): 175–192.

Pauly, D., and Tsukayama, I. (*Editors*). 1987. The Peruvian anchoveta and its upwelling ecosystem: three decades of change. International Center for Living Aquatic Resources Management (ICLARM), Manila, Phillipines. 351 p.

Pauly, D., Christensen, V., Dalsgaard, J., Froese, R., and Torres, F.C., Jr. 1998. Fishing down marine food webs. Science (Washington, D.C.) **279**: 860–863.

Pearce, David, Markandya, Anil, and Barbier, Edward B. 1989. Blueprint for a green economy. Earthscan, London.

Pearcy, W.G. 1992. Ocean ecology of north Pacific salmonids. Washington Sea Grant Program, University of Washington, Seattle, WA. 179 p.

Pearse, Peter. 1982. Turning the tide: the commission on Pacific fisheries policy. Canada Department of Fisheries and Oceans, Pacific Region, Vancouver, BC.

Pearse, P.H., and Wilen, J.E. 1979. Impact of Canada's Pacific salmon fleet control program. J. Fish. Res. Board Can. **36**: 764–769.

Perry, R.I., Walters, C.J., and Boutillier, J.A. 1999. A framework for providing scientific advice for the management of new and developing invertebrate fisheries. Rev. Fish Biol. Fish. **9**: 125–150.

Peterman, R.M. 1977. A simple mechanism that causes collapsing stability regions in exploited salmonid populations. J. Fish. Res. Board Can. **34**: 1130–1142.

Peterman, R.M. 1984. Density-dependent growth in early ocean life of sockeye salmon (*Oncorhynchus nerka*). Can. J. Fish. Aquat. Sci. **41**(12): 1825–1829.

Peterman, R.M. 1987. Review of the components of recruitment of Pacific salmon. *In* Common strategies of anadromous and catadromous fishes. *Edited by* M. Dadswell, R. Klauda, C. Moffitt, R. Saunders, R. Rulifson, and J.E. Cooper. Am. Fish Soc. Symp. **1**: 417–429.

Peterman, R.M. 1990. Statistical power analysis can improve fisheries research and management. Can. J. Fish. Aquat. Sci. **47**: 2–15.

Peterman, R.M., and Bradford, M.J. 1987. Wind speed and mortality rate of a marine fish, the northern anchovy *(Engraulis mordax)*. Science (Washington, D.C.) **235**: 354–356.

Peterman, R.M., Pyper, B.J., Lapointe, M.F., Adkison, M.D., and Walters, C.J. 1998*a*. Patterns of covariation in survival rates of British Columbian and Alaskan sockeye salmon stocks. Can. J. Fish. Aquat. Sci. **55**: 2503–2517.

Peterman, R.M., Peters, C.N., Robb, C.A., Frederick, S.W. 1998*b*. Bayesian decision analysis and uncertainty in fisheries management. *In* Reinventing fisheries management. *Edited by* T.J. Pitcher, P.J.B. Hart, and D. Pauly. Kluwer Academic Publishers, London, U.K. pp. 387–398.

Peterson, S., and Smith, L.J. 1982. Risk reduction in fisheries management. Ocean. Manage. **8**: 65–79.

Philander, S.G. 1990. El Nino, La Nina and the Southern Oscillation. Academic Press, San Diego, CA. 293 p.

Phillips, B.F., Pearce, A.F., Litchfield, R., and Guzman Del Proo, S.A. 1994*a*. Spiny lobster catches and the ocean environment. *In* Spiny lobster management. *Edited by* B.F. Phillips, J.S. Cobb, and J. Kittaka. Blackwell Scientific Pub., Oxford. pp. 250–261.

Phillips, B.F., Cruz, R., Brown, R.S., and Caputi, N. 1994*b*. Predicting the catch of spiny lobster fisheries. *In* Spiny lobster management. *Edited by* B.F. Phillips, J.S. Cobb, and J. Kittaka. Blackwell Scientific Pub., Oxford. pp. 285–301.

Pigou, A.C. 1920. The economics of welfare. Macmillan, London.

Pimm, S.L. 1997. The value of everything. Nature (Lond.) **387**: 261–264.

Pinkerton, E. 1987. Intercepting the state: dramatic processes in the assertion of local comanagement rights. *In* The question of the commons: the culture and ecology of communal resources. *Edited by* B.J. McCay and J. Acheson. Univ. Arizona Press, Tucson, AZ. pp. 344–369.

Pinkerton, E.W. 1989. Introduction: attaining better fisheries managment through co-management — prospects, problems and propositions. *In* Co-operative management of local fisheries: new directions for improved management and community development. *Edited by* E.W. Pinkerton. UBC Press, Vancouver, BC.

Pinkerton, E. 1994. Economic and management benefits from the coordination of capture and culture fisheries: the case of Prince William Sound pink salmon. N. Am. J. Fish. Manage. **14**: 262–277.

Pinkerton, E., and Weinstein, M. 1995. Fisheries that work: sustainability through community-based management. The David Suzuki Foundation,Vancouver, BC. 199 p.

Pogson, G.H., Mesa, K.A., and Boutillier, R.G. 1995. Genetic population structure and gene flow in Atlantic cod *Gadus morhua*: a comparison of allozyme and nuclear RFLP loci. Genetics **139**: 375–385.

Policansky, D. 1993. Evolution and management of exploited fish populations. *In* Proceedings of the International Symposium on Management Strategies for Exploited Fish Populations. *Edited by* G. Kruse, D.M. Eggers, R.J. Marasco, C. Pautzke, T.J. Quinn II. Alaska Sea Grant College Program Report No. 93-02, University of Alaska, Fairbanks, AL. pp. 651–664.

Prena, J., Rowell, T.W., Schwinghamer, P., Gilkinson, K., and Gordon, D.C., Jr. 1996. Grand Banks otter trawling impact experiment: 1. Site selection process, with a description of macrofaunal communities. Can. Tech. Rep. Fish. Aquatic Sci. 2094. 38 p.

Pringle, J.D., and Burke, D.L. 1993. The Canadian lobster fishery and its management, with emphasis on the Scotian Shelf and the Gulf of Maine. *In* Perspectives on Canadian marine fisheries management. *Edited by* L.S. Parsons and W.H. Lear. Can. Bull. Fish. Aquat. Sci. 226. pp. 91–122.

Punt, A.E., and Hilborn, R. 1997. Fisheries stock assessment and decision analysis: the Bayesian approach. Rev. Fish Biol. Fish. **7**: 35–63.

Pyper, B.J., Peterman, R.M., Lapointe, M.F., Walters, C.J. 1999. Patterns of covariation in length and age at maturity of British Columbia and Alaska sockeye salmon (*Oncorhynchus nerka*) stocks. Can. J. Fish. Aquat. Sci. **56**: 1046–1057.

Raiffa, H. 1982. The art and science of negotiation. Harvard University Press, Cambridge, MA.

Repetto, R., Magrath, W., Wells, M., Beer, C., and Rossini, F. 1989. Wasting assets: natural resources in the national accounts. World Resources Institute, Washington, D.C.

Restrepo, V.R., Hoenig, J.M., Powers, J.E., Baird, J.W., and Turner, S.C. 1992. A simple simulation approach to risk and cost analysis, with applications to swordfish and cod fisheries. Fish. Bull. (U.S.) **90**: 736–748.

Richards, L.J., and Maquire, J.-J. 1998. Recent international agreements and the precautionary approach: new directions for fisheries management science. Can. J. Fish. Aquat. Sci. **55**: 1545–1552.

Ricker, W.E. 1954. Stock and recruitment. J. Fish. Res. Board Can. **11**: 559–623.

Ricker, W.E. 1975. Computation and interpretation of biological statistics of fish populations. Bull. Fish. Res. Board Can. 191. 382 p.

Rivard, D., and Maguire, J.J. 1993. Reference points for fisheries management: the eastern Canadian experience. *In* Risk evaluation and biological reference points for fisheries management. *Edited by* S.J. Smith, J.J. Hunt, and D. Rivard. Can. Spec. Publ. Fish Aquat. Sci. 120. pp. 31–58.

Rogers, D.E., and Ruggerone, G.T. 1993. Factors affecting marine growth of Bristol Bay sockeye salmon. Fish. Res. **18**: 89–103.

Rose, G.A., and Leggett, W.C. 1988. Atmospheric-ocean coupling and Atlantic cod migrations: effects of wind-forced variations in sea temperatures and currents on nearshore distributions and catch rates of *Gadus morhua*. Can. J. Fish. Aquat. Sci. **45**: 1234–1243.

Rose, G.A. 1993. Cod spawning on a migration highway in the north-west Atlantic. Nature (Lond.) **366**: 458–461.

Rosenberg, A.A., and Brault, S. 1993. Choosing a management strategy for stock rebuilding when control is uncertain. Can. Spec. Publ. Fish. Aquat. Sci. 120. pp. 243–252.

Rosenberg, A.A., and Restrepo, V.R. 1994. Uncertainty and risk evaluation in stock assessment advice for U.S. marine fisheries. Can. J. Fish. Aquat. Sci. **51**: 2715–2720.

Rosenberg, A.A., Fogarty, M.J., Sissenwine, M.P., Beddington, J.R., and Shepherd, J.G. 1993. Achieving sustainable use of renewable resources. Science (Washington, D.C.) **262**: 828–829.

Rothschild, B.J. 1986. Dynamics of marine fish populations. Harvard Univ. Press, Cambridge, MA.

Rothschild, B.J., and Heimbuch, D. 1983. Managing variable fishery stocks in an uncertain environment: the design of fishery management systems. FAO Tech. Rept. 291 Vol. 3. pp. 1141–1160.

Rothschild, B.R., and Osborn, T.R. 1988. Small-scale turbulence and plankton encounter rates. J. Plank. Res. **10**: 465–474.

Rowan, K. 1994. Why rules for risk communication are not enough: a problem solving approach to risk communication. Risk Anal. **14**(3): 365–374.

Ryder, R.A, and Scott, W.B. 1994. Effects of fishing on biodiversity in Canadian waters. *In* Biodiversity in Canada: a science assessment for Environment Canada. *Edited by* Biodiversity Science Team. Environment Canada, Ottawa, Canada. 245 p.

Ryman, N. 1991. Conservation genetics considerations in fishery management. J. Fish. Biol. **39**(Suppl. A): 211–224.

Ryman, N., Utter, F., and Laikre, L. 1995. Protection of intraspecific biodiversity of exploited fishes. Rev. Fish Biol. Fish. **5**: 417–446.

Saetersdal, G. 1980. A review of past management of some pelagic stocks and its effectiveness. Rapp. P.-V. Reun. Cons. Int. Explor. Mer **177**: 505–512.

Sagoff, Mark. 1997. Can we put a price on nature's services? Philosophy and public policy, summer 1997. Institute for Philosophy and Public Policy, University of Maryland, College Park, MD.

Sainsbury, K.J., Campbell, R.A., Lindholm, R., and Whitelaw, A.W. 1997. Experimental management of an Australian multispecies fishery: examining the possibility of trawl-induced habitat modification. *In* Global trends: fisheries management. *Edited by* E.L. Pikitch, D.D. Huppert, and M.P. Sissenwine. Bethesda, MD. Am. Fish. Soc. Symp. **20**: 107–112.

Savard, L. (*Editor*). 1994. Status report on invertebrates 1993: crustaceans and molluscs on the Quebec coast and northern shrimp in the estuary and Gulf of St. Lawrence. Can. Manuscr. Rep. Fish. Aquat. Sci. 2257. 113 p.

Savard, J-P. L., Freemark, K., and Reynoldson, T. 1994. General concepts related to biodiversity. *In* Biodiversity in Canada: a science assessment for Environment Canada. *Edited by* Biodiversity Science Team. Environment Canada, Ottawa, Canada. 245 p.

Schaefer, M.B. 1954. Some aspects of the dynamics of populations important to the management of the commercial marine fisheries. Bull. Int.-Am. Trop. Tuna Comm. **1**: 27–56.

Schlager, E., and Ostrom, E. 1993. Property-rights regimes and coastal fisheries: an empirical analysis. *In* The political economy of customs and culture: informal solutions to the commons problem. *Edited by* T.L. Anderson and R.T. Simmons. Rowman and Littlefield, Lanham, MD. pp. 13–42.

Schopka, S.A. 1994. Fluctuations in the cod stock off Iceland during the 20th century in relation to changes in fisheries and environment. ICES Mar. Sci. Symp. **198**: 175–193.

Schrank, W.E. 1996. Origins of Atlantic Canada's fisheries crisis. *In* Fisheries and uncertainty: a precautionary approach to resource management. *Edited by* D.V. Gordon and G.R. Munro. Univ. Calgary Press, Calgary, AB. pp. 27–44.

Scientific Panel for Sustainable Forest Practices in Clayoquot Sound. 1995. Report. Cortex Consultants, Victoria, BC.

Sened, I. 1997. The political institution of private property. Cambridge University Press, Cambridge, MA.

Serchuk, F.M., Grosslein, M.D., Lough, R.G., Mountain, D.G., and O'Brien, L. 1994. Fishery and environmental factors affecting trends and fluctuations in the Georges Bank and Gulf of Maine Atlantic cod stocks: an overview. ICES Mar. Sci. Symp. **198**: 77–109.

Shelton, P.A., and Atkinson, D.B. 1994. Proceedings of the regional groundfish assessment review for Newfoundland, May 9–13, 1994. Can. Tech. Rep. Fish. Aquat. Sci. 2020. 100 p.

Shelton, P.A., and Morgan, M.J. 1993. Assessing the risk of failing to achieve replacement recruitment. ICES CM 1993/D:54. 20 p.

Shelton, P.A., Stansbury, D.E., Murphy, E.F., Lilly, G.R., and Brattey, J. 1996. An assessment of the cod stock in NAFO divisions 2J+3KL. DFO Atl. Fish. Res. Doc. 96/80. 65 p.

Shepherd, J.G. (*Editor*). 1991. Special session on management under uncertainties. Sci. Counc. Simon, Herbert. 1947. Administrative behavior. MacMillan, New York.

Sinclair, M. 1988. Marine fish populations. Univ. of Washington Press, Seattle, WA.

Sinclair, P., and Felt, L. 1993. Coming back: return migration to Newfoundland's Great Northern Peninsula. Newfoundland Stud. **9**(1): 1–25.

Sinclair, A., Chouinard, G., and Currie, L. 1996. Assessment of the southern Gulf of St. Lawrence cod stock, March 1996. Stud. 16. Northwest Atl. Fish. Organ., Dartmouth, NS. DFO Atl. Fish. Res. Doc. 96/52. 75 p.

Sissenwine, M.P. 1986. Perturbation of a predator-controlled ecosystem. *In* Variability and management of large marine ecosystems. *Edited by* K. Sherman and L. Alexander. AAAS Selected Symp. 99. Westview Press, Boulder, CO. pp. 55–85.

Sissenwine, M.P., and Cohen, E.B. 1991. Resource productivity and fisheries management: Northeast Shelf ecosystem. *In* Food chains, yields, models and management of large marine ecosystems. *Edited by* K. Sherman and G. Gold. AAAS Selected Symp. 99 Westview Press, Boulder, CO. pp. 107–123.

Sissenwine, M.P., Azarovitz, T.R., and Suomala, J.B. 1983. Determining the abundance of fish. *In* Experimental biology at sea. *Edited by* A.G. MacDonald and I.G. Priede. Academic Press, London. pp. 51–101.

Sivasubramaniam, K. 1993. The biosocioeconomic way: a new approach to management of small-scale fisheries in the Bay of Bengal region. Bay of Bengal News **52**: 4–15.

Slaney, T.L., Hyatt, K.D., Northcote, T.G., and Fielden, R.J. 1996. Status of anadromous salmon and trout in British Columbia and Yukon. Fisheries **21**(10): 20–35.

Slocombe, D.S. 1993. Environmental planning, ecosystem science and ecosystem approaches for integrating environment and development. Environ. Manage. **17**: 289–298.

Slovic, P. 1987. Perception of risk. Science (Washington, D.C.) **230**: 280–285.

Smayda, T. 1990. Novel and nuisance phytoplankton blooms in the sea: evidence for a global epidemic. *In* Toxic marine phytoplankton. *Edited by* E. Graneli, B. Sundstrom, L. Edler and D.M. Anderson. Elsevier, New York. pp. 29–40.

Smed, J. 1980. Temperature of the waters off southwest and south Greenland during the ICES/ICNAF salmon tagging experiment in 1972. Rapp. P.-V. Reun. Cons. Int. Explor. Mer **176**: 18–21.

Smith, Adam. 1776 [1976]. An inquiry into the nature and causes of the wealth of nations. 1976 edition. Clarendon Press, Oxford.

Smith, I.R. 1981. Improving fishing incomes when resources are overfished. Mar. Policy **5**: 17–22.

Smith, P.J., Francis, R.I.C.C., and McVeagh, M. 1991. Loss of genetic diversity due to fishing pressure. Fish. Res. **10**: 309–316.

Smith, S.J., Hunt, J.J., and Rivard, D. (*Editors*). 1993. Risk evaluation and biological reference points for fisheries management. Can. Spec. Publ. Fish. Aquat. Sci. 120.

Smith, V.L. 1968. Economics of production from natural resources. Am. Econ. Rev. **58**: 409–431.

Solomon, S. 1990. Progress towards a quantitative understanding of Antarctic ozone depletion. Nature (Lond.) **347**: 347–354.

Sparrow - Regina v. Sparrow, Supreme Court of Canada, May 31, 1990.

Stamatopoulos, C. 1993. Trends in catches and landings. Atlantic fisheries: 1970–1991. Food and Agriculture Organization (FAO). Fish Circ. 855.1. FAO, Rome. 223 p.

Stanistreet, K. 1982. Limited entry in the abalone fishery of Victoria. *In* Policy and practice in fisheries management. Proceedings of the National Fisheries Seminar on Economic Aspects of Limited Entry and Associated Fisheries Management Measures, Australian Government Publishing Service, Canberra. pp. 139–152.

Statistics Canada. 1997*a*. Econnections: linking the environment and the economy. Concepts, sources and methods of the Canadian system of environmental and resource accounts. Cat 16-505-GPE. Ottawa, Canada.

Statistics Canada 1997*b*. Econnections: linking the environment and the economy. Indicators and detailed statistics. Cat 16-200-XKE. Ottawa, Canada.

Steele, J.H. 1985. Comparison of marine and terrestrial ecological systems. Nature (Lond.) **313**: 355–358.

Steele, J.H., and Henderson, E.W. 1984. Modeling long-term fluctuations in fish stocks. Science (Washington, D.C.) **224**: 985–987.

Stephenson, R.L., and Lane, D.E. 1995. Fisheries science in fisheries management: a plea for conceptual change. Can. J. Fish. Aquat. Sci. **52**: 2051–2056.

Stommel, H. 1957. The abyssal circulation of the ocean. Nature (Lond.) **180**: 733–734.

Taylor, A.H. 1995. North-south shifts of the Gulf Stream and their climatic connection with the abundance of zooplankton in the UK and the surrounding seas. ICES J. Mar. Sci. **52**: 711–721.

Sutcliffe, W.H., Jr. 1973. Correlations between seasonal river discharge and local landings of American lobster (*Homarus americanus*) and Atlantic halibut (*Hippoglossus hippoglossus*) in the Gulf of St. Lawrence. J. Fish. Res. Board Can. **34**: 19–30.

Taggart, C.T. 1997. Bank-scale migration patterns in northern cod. NAFO Sci. Coun. Studies. **29**: 51–60.

Taggart, C.T., Anderson, J., Bishop, C., Colbourne, E., Hutchings, J., Lilly, G., Morgan, J., Murphy, E., Myers, R., Rose, G., and Shelton, P. 1994. Overview of cod stocks, biology, and environment in the Northwest Atlantic region of Newfoundland, with emphasis on Northern Cod. ICES Mar. Sci. Symp. **198**: 140–157.

Taggart, C.T., Penney, P., Barrowman, N., and George, C. 1995. The 1954–1993 Newfoundland cod-tagging data base: statistical summaries and spatial-temporal distributions. Can. Tech. Rep. Fish. Aquat. Sci. 2042. 441 p. + appendices.

Templeman, W. 1979. Migration and intermingling of stocks of Atlantic cod, *Gadus morhua*, of the Newfoundland and adjacent areas from tagging in 1962–66. Int. Comm. NW Fish. Res. Bull. **14**. 50 p.

Thompson, K.R., and Page, F.H. 1989. Detecting synchrony of recruitment using short, autocorrelated time series. Can. J. Fish. Aquat. Sci. **46**: 1831–1838.

Trenberth, K.E., and Hurrell, J.W. 1995. Decadal coupled atmosphere-ocean variations in the North Pacific ocean. *In* Climate change and northern fish populations. *Edited by* R.J. Beamish. Can. Spec. Publ. Fish. Aquat. Sci. 121. pp. 15–24.

Trippel, E.A. 1995. Age at maturity as a stress indicator in fisheries. BioScience **45**: 759–771.

Trippel, E.A. 1998. Egg size, viability and seasonal offspring production of young Atlantic cod. Trans. Am. Fish. Soc. (In press).

United Nations (UN). 1993. SNA Handbook on integrated environmental and economic accounting. Statistical office of the United Nations, Series F, No 61, New York.

United Nations (UN) . 1995. Agreement for the implementation of the provisions of the United Nations Convention on the law of the sea of 10 December 1982 relating to the conservation and management of straddling fish stocks and highly migratory fish stocks. A/Conf 164/37 (8 September 1995). United Nations, New York.

Upton, H.F. 1992. Biodiversity and conservation of the marine environment. Fisheries **17**(3): 20–25.

van Dieren, Wouter. 1995. Taking nature into account. Springer-Verlag, New York.

VenTresca, D.A., Parrish, R.H., Houk, J.L., Gingras, M.L., Short, S.D., and Crane, N.L. 1995. El Nino effects on the somatic and reproductive condition of blue rockfish, *Sebastes mystinus*. CalCOFI Rep. 36. pp. 167–174.

von Weizsäcker, Ernst Ulrich. 1994. Earth politics. Zed Books, London.

Voytek, M.A. 1990. Addressing the biological effects of decreased ozone on the Antarctic environment. Ambio **19**: 52–61.

Wainwright, S.C., Fogarty, M.J., Greenfield, R.C., and Fry, B. 1993. Long-term changes in the Georges Bank food web: trends in stable isotopic compositions of fish scales. Mar. Biol. **115**: 481–493.

Walters, C.J. 1981. Optimum escapements in the face of alternative recruitment hypotheses. Can. J. Fish. Aquat. Sci. **38**(6): 678–689.

Walters, C.J. 1986. Adaptive management of renewable resources. MacMillan, New York. 374 p.

Walters, C.J. 1987. Nonstationarity of production relationships in exploited fish populations. Can. J. Fish. Aquat. Sci. **44**(Suppl. 2): 156–165.

Walters, C.J. 1995. Fish on the line: the future of Pacific fisheries. A report to the David Suzuki Foundation Fisheries Project, Phase I, Vancouver, BC.

Walters, C.J., and Collie, J.C. 1988. Is research on environmental factors useful to fisheries management? Can. J. Fish. Aquat. Sci. **45**: 1848–1854.

Ware, D.M. 1991. Climate, predators and prey: behavior of a linked oscillating system. pp. 279–291. *In* Long-term variability in pelagic fish populations and their environment. *Edited by* T. Kawasaki, S. Tanaka, Y. Toba, and A. Taniguchi. Proceedings of the International

Symposium, 14–18 November 1989, Sendai, Japan. Pergamon Press, Tokyo, Japan.

Ware, D.M., and Thomson, R.E. 1991. Link between long-term variability in upwelling and fish production in the northeast Pacific ocean. Can. J. Fish. Aquat. Sci. **48**: 2296–2306.

Waring, Marilyn. 1988. If women counted. Harper and Row, San Francisco, CA.

Welch, D.W., Chigrinsky, A.I., and Ishida, Y. 1995. Upper thermal limits on the oceanic distribution of Pacific salmon (*Oncorhynchus spp.*) in the spring. Can. J. Fish. Aquat. Sci. **52**: 489–503.

Wells, P.G., and Ricketts, P.J. 1994. Coastal zone Canada '94, cooperation in the coastal zone: conference proceedings. Coastal Zone Canada Association, Bedford Institute of Oceanography, Dartmouth, NS.

Wiebe, P.H., and Beardsley, R.C. 1996. Introduction: physical–biological interactions on Georges Bank and its environs. Deep-Sea Res. (Part II) **43**: 1437–1438.

Winter, B.D., and Hughes, R.M. 1995. AFS draft position statement on biodiversity. Fisheries **20**(4): 20–26.

Winter, B.D., and Hughes, R.M. 1997. Biodiversity position statement — corrected version. Fisheries **22**(3): 16–23.

Woodey, J.C. 1987. In-season management of Fraser River sockeye salmon: meeting multiple objectives. *In* Sockeye salmon (*Oncorhynchus nerka*) population biology and future management. *Edited by* H.D. Smith, L. Margolis, and C.C. Wood. Can. Spec. Publ. Fish. Aquat. Sci. 96. pp. 367–374.

Wooster, W.S., and Fluharty, D.L. (*Editors*). 1985. El Nino North: Nino effects in the eastern subarctic Pacific Ocean. Washington Sea Grant Program, University of Washington, Seattle, WA.

Wynne, B.1992. Uncertainty and environmental learning: reconceiving science and policy in the preventive paradigm. Global Environ. Change **2**(2): 111–127.